# SpringerBriefs in Applied Sciences and Technology

T0172038

SpringerBriefs present concise summaries of cutting-edge research and practical applications across a wide spectrum of fields. Featuring compact volumes of 50 to 125 pages, the series covers a range of content from professional to academic.

Typical publications can be:

- A timely report of state-of-the art methods
- An introduction to or a manual for the application of mathematical or computer techniques
- A bridge between new research results, as published in journal articles
- A snapshot of a hot or emerging topic
- An in-depth case study
- A presentation of core concepts that students must understand in order to make independent contributions

SpringerBriefs are characterized by fast, global electronic dissemination, standard publishing contracts, standardized manuscript preparation and formatting guidelines, and expedited production schedules.

On the one hand, **SpringerBriefs in Applied Sciences and Technology** are devoted to the publication of fundamentals and applications within the different classical engineering disciplines as well as in interdisciplinary fields that recently emerged between these areas. On the other hand, as the boundary separating fundamental research and applied technology is more and more dissolving, this series is particularly open to trans-disciplinary topics between fundamental science and engineering.

Indexed by EI-Compendex, SCOPUS and Springerlink.

More information about this series at http://www.springer.com/series/8884

Max F. Platzer • Nesrin Sarigul-Klijn

# The Green Energy Ship Concept

## Renewable Energy from Wind Over Water

Max F. Platzer
Innovative Power Generation
Systems (iPGS) Laboratory,
Department of Mechanical
and Aerospace Engineering
University of California, Davis
Davis, CA, USA

Nesrin Sarigul-Klijn
Innovative Power Generation
Systems (iPGS) Laboratory,
Department of Mechanical
and Aerospace Engineering
University of California, Davis
Davis, CA, USA

ISSN 2191-530X          ISSN 2191-5318    (electronic)
SpringerBriefs in Applied Sciences and Technology
ISBN 978-3-030-58243-2        ISBN 978-3-030-58244-9    (eBook)
https://doi.org/10.1007/978-3-030-58244-9

This Springer imprint is published by the registered company Springer Nature Switzerland AG
The registered company address is: Gewerbestrasse 11, 6330 Cham, Switzerland

# Preface

We are passengers in a spaceship that is getting warmer year after year. The climatologists tell us that the continued emission of greenhouse gases will risk the onset of irreversible climate change by mid-century. Many influential people have been warning about the consequences, and many people, especially the young ones, have been expressing their anger and frustration about the lack of an effective response. In his recent addresses to the United Nations General Assembly in New York in September 2019 and the United Nations COP25 conference in Madrid in December 2019, the United Nations Secretary General Antonio Guterres stressed the need for the declaration of a global climate emergency. Unfortunately, COP25 yielded no consensus on how to respond.

The aim of our book is to draw attention to the fact that the record of past engineering achievements should encourage us to meet this century's challenge of developing sustainable energy conversion systems. Only 120 years ago, it was unimaginable that engineering would make it possible to "conquer space and time." By this, we mean the development of airplanes, which enabled the transport of hundreds of people in a single plane to any point on the globe within a day, and the development of rockets, which in turn made possible instant communication between people on any place on the globe.

This raises the question of whether the lack of a consensus on how to respond to the UN declaration of emergency is due to disagreement about the most effective engineering solutions; or merely squabbling about the distribution of the necessary sacrifices among the various nations. It is our view that the former question needs to be answered before there can be any hope to achieve a global sociopolitical consensus.

Given the huge global energy demand to maintain our standard of living, there are only three renewable energy sources that can satisfy this demand, namely, solar, water, and wind. The development of the global hydropower plants started already in the late nineteenth century and therefore leaves little room for further substantial contributions. The major contributions will have to come from solar and wind power plants. Indeed, the solar and wind power engineering industries have experienced impressive growth in the past half-century. To wit, wind turbine blades are now longer than the largest airplane wings. However, there is a generally accepted view that solar panels and solar and wind power plants require a solid foundation and therefore need to be land-based or at most offshore based in shallow coastal waters. This view eliminates two wind energy sources from consideration, namely, jet streams and winds over the oceans.

Aerospace engineers have a tendency to look for challenging problems in their field. In 2009 we published a proposal on how to capture the wind energy in both the jet streams and the winds over the oceans. We soon realized that the jet stream capture is too challenging and therefore concentrated on the ocean wind capture problem. We approached it with the attitude that we would have to show technical feasibility before worrying about costs. To our surprise, we found that the technical elements needed to convert wind over water into storable energy were readily available, although they certainly required further development and identification of the optimum combination of the many parameters that have an influence. We were fortunate in attracting the interest of Professor Peter Pelz at the Technical University of Darmstadt, Germany. He and his Ph.D. student Mario Holl subjected our energy ship concept to a techno-economic multipole systems analysis. It revealed not only the optimal extractable energy as a function of wind speed, wind angle, sail area, ship and turbine drag, electrolyzer efficiency, etc., but also provided a hydrogen production cost estimate. As expected, it showed the need for minimizing the ship drag and the personnel costs, thus pointing to the operation of autonomous hydrofoil boats as the most effective energy ship.

It is our view that enough information is now available on our "wind-over-water" energy conversion concept to summarize its present development status in this book for the purpose of making it accessible to a wider readership. To this end, we divided it into two parts. In the first part, we briefly summarize the nature of the climate crisis and its potential irreversibility, the current status of renewable energy technologies and efforts, as well as our reasons for proposing the energy ship concept. In the second part, we provide brief summaries of the essential technologies needed for the implementation of the concept. These summaries are merely meant as "appetizers" for the technically interested reader to stimulate more detailed study by consulting the listed references.

In our view, the lack of a consensus at COP25 is partly due to the lack of appreciation for the opportunities offered by the exploitation of the wind energy over the oceans. It requires a combination of aeronautical, hydronautical, and power engineering to develop efficient air-sea interface vehicles. We like to call this new engineering discipline, "aero-hydronautical power engineering." It is our great hope that this new discipline can make a significant contribution to the alleviation of the climate emergency declared by the UN Secretary General.

This book evolved from lectures we gave every year over the past ten years in a seminar course to first-year UC Davis students for the purpose of making them aware of the challenges presented by the climate change and of potential engineering solutions to these challenges. During the past 2 years, 20 senior design students worked on renewable energy design projects in our innovative Power Generation

Systems (iPGS) laboratory. Also, one of us (MFP) benefited greatly from the feedback he received during lectures at the Royal Institute of Technology in Stockholm, the Graz University of Technology, the Istanbul Technical University, the Eskisehir Technical University, the University of Stuttgart, the Tsinghua University of Beijing, and the Hong Kong University of Science and Technology. He expresses his gratitude for the hospitality extended by Professors Fransson, Jericha, Sanz, Unal, Karakoc, Vogt, Song Fu, and Wei Shyy at these universities. He is also grateful for stimulating discussions with Professors Hobson and Gannon at the Naval Postgraduate School and Professor Turner at the University of Cincinnati.

Davis, CA, USA                                                   Max F. Platzer
                                                          Nesrin Sarigul-Klijn

# Contents

# About the Author

**Max F. Platzer** is an Adjunct Professor of Mechanical and Aerospace Engineering at the University of California, Davis. He holds Diplom-Ingenieur and Doctor of Technical Sciences degrees from the Technical University of Vienna, Austria. He was a member of Wernher von Braun's SATURN rocket development team for 6 years, chief of the Aeromechanics Research Section at the Lockheed-Georgia Research Center for 4 years, and a Professor of Aeronautics and Astronautics at the Naval Postgraduate School, Monterey, California, for 34 years. Dr. Platzer received the distinguished professor medal of the Naval Postgraduate School. He is a Fellow of the American Institute of Aeronautics and Astronautics and of the American Society of Mechanical Engineers. Currently, he is editor of the international review journal *Progress in Aerospace Sciences*.

**Nesrin Sarigul-Klijn** is a Full Professor of Mechanical and Aerospace Engineering and the Founding Director of the Space Engineering Research and Graduate Program at the University of California, Davis. She received her Ph.D. degree from the University of Arizona. In addition to her engineering academic degrees, she is an instrument-rated commercial pilot and an active participant of FAA wings. Her publications' record of over 200 refereed technical works also includes five patents and two books. She serves on the Editorial Board of the journal *Progress in Aerospace Sciences*. She is an Associate Fellow of the American Institute of Aeronautics and Astronautics and Fellow of the American Society of Mechanical Engineers. Her cross-disciplinary research expertise is in fluid-structure interactions, acoustics and noise control, vibrations, dynamic separation of air-launched vehicles, and autonomous systems.

# Part I
# General Considerations

# Chapter 1
# Introduction

Currently, about 80 percent of the global energy demand is satisfied by fossil-based energy production methods which emit harmful greenhouse gases. The reduction and eventual elimination of these gases therefore has been recognized as an increasingly urgent goal if irreversible climate change is to be avoided. There is much doubt, however, whether a complete transition of the global economy from fossil-based to renewable energy production can be achieved. Most studies project this transition to occur gradually over the remaining decades of this century. However, there is a growing realization that the transition needs to be completed by no later than 2050 if irreversible climate change is to be averted. This therefore raises the question whether an Apollo-type engineering program is possible to accomplish this objective within 30 years. This then is the challenge for the global engineering community which we propose to discuss in this book.

It is our view that this challenge requires the participation of the global engineering community in the form of a Global Renewable Energy Engineering Council consisting of highly qualified representatives of the major national engineering societies. It shall be the task of this Council to set aside any consideration of the socio-political feasibility aspects of this Global Energy Apollo Program and to limit its deliberations exclusively to the pure engineering feasibility aspects.

We therefore attempt to review in this book the current status of global renewable energy production and storage initiatives and provide an assessment of their potential to reach the transition goal by no later than 2050. We then proceed to argue that the current engineering initiatives ignore the exploitation of the vast wind resources available in the form of the winds over the oceans. We refer to the dramatic engineering breakthroughs achieved in World War II and during the Cold War as examples that give us the confidence to argue that equally innovative engineering solutions to the climate crisis are achievable within 30 years. The remaining chapters of our book therefore are devoted to the elucidation of this possibility, supported by additional technical information provided in Part II.

© The Editor(s) (if applicable) and The Author(s), under exclusive license to
Springer Nature Switzerland AG 2021
M. F. Platzer, N. Sarigul-Klijn, *The Green Energy Ship Concept*, SpringerBriefs in
Applied Sciences and Technology, https://doi.org/10.1007/978-3-030-58244-9_1

In the final chapter we emphasize the need for an objective technical evaluation of the engineering feasibility and economic cost of the potential engineering approaches to meet the goal of the Global Energy Apollo Program so that the various national governments and their citizens can rely on the answer by the world's most respected engineering experts as to the timeline and cost of a global program to achieve an emission-free global economy.

We start this book with the stipulation that climate change is an anthropogenic phenomenon caused by the introduction of the steam engine for power generation in the eighteenth century, followed by the invention of the steam turbine and the Otto and Diesel motors in the nineteenth century and of the gas turbine in the twentieth century. Their use has made it possible to provide the average citizen of the developed countries with a standard of living unimaginable only two centuries ago. Unfortunately, most citizens became aware only very recently of the harmful effect of continued emission of greenhouse gases on the climate although Svante Arrhenius explained already in 1896 the influence of heat-absorbing gases in the atmosphere on the mean temperature on the ground [1].

There is now a general consensus among atmospheric scientists that global warming due to the emission of greenhouse gases from various anthropogenic activities is a real phenomenon. Many papers and books have been published in recent years to warn about its potentially catastrophic effect on future generations due to the fact that the carbon dioxide emissions stay in the atmosphere for many hundreds of years. The continued addition of carbon dioxide therefore makes it highly likely that our planet will cross several tipping points within the coming decades which will trigger irreversible changes [2]. This possibility has been carefully documented in the latest paper published in the proceedings of the US Academy of Sciences [3]. A number of authors have tried for years to warn the general public that in a few decades the average temperatures will rise to levels which will render certain parts of our planet uninhabitable, for example, Hansen [4] and Gore [5]. In a very recent issue, The New York Times Magazine [6] warns its readers that we risk the collapse of civilization if we do not reduce the greenhouse gas emissions.

We limit ourselves to the presentation of only three figures which we regard as entirely sufficient to demonstrate the anthropogenic cause of climate change and its adverse effect on human civilization.

Figure 1.1 shows the dramatic increase of the world population in the past 200 years from well under 1 billion in the early 1800s to over 7 billion at the present time and a projected population of over 9 billion by the end of the century. Figure 1.2 shows the equally dramatic increase of the measured $CO_2$ emission, while Fig. 1.3 depicts the measured carbon dioxide increase in the atmosphere. The accuracy of these data is not in doubt making it difficult to deny the strong correlation in trends between these figures.

Nevertheless, there are many people who are ignoring or even denying the anthropogenic cause of the sea level rise. This is quite understandable because many people do not have the time and/or interest in informing themselves about a phenomenon which appears to be complex, controversial, and, most importantly, of

## World Population Growth Through History

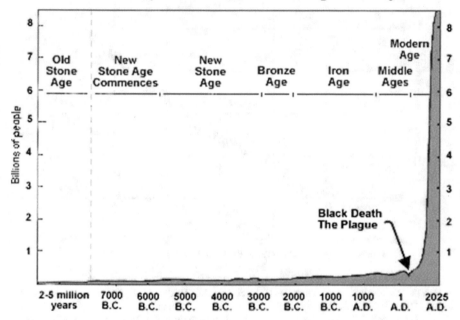

**Fig. 1.1** World population growth
Source: Population Reference Bureau

## Global CO₂ emissions by world region, 1751 to 2015

Annual carbon dioxide emissions in billion tonnes (Gt).

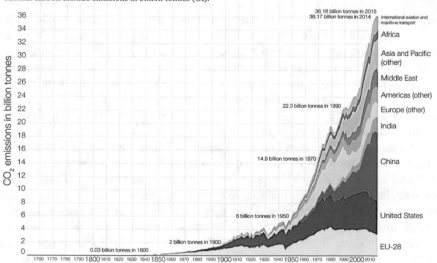

**Fig. 1.2** Carbon dioxide emission versus time

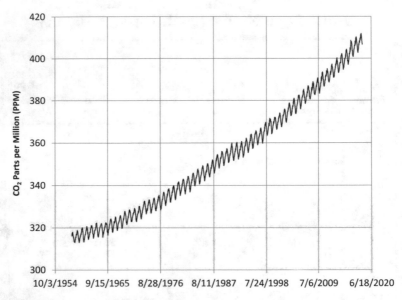

**Fig. 1.3** Atmospheric carbon dioxide emission growth
Source: C. D. Keeling, S. C. Piper, R. B. Bacastow, M. Wahlen, T. P. Whorf, M. Heimann, and H. A. Meijer, Exchanges of atmospheric CO2 and 13CO2 with the terrestrial biosphere and oceans from 1978 to 2000. I. Global aspects, SIO Reference Series, No. 01–06, Scripps Institution of Oceanography, San Diego, 88 pages, 2001

no immediate threat to their standard of living. It is this absence of an immediate threat which presents the greatest challenge to mobilizing the general public to demand a rapid reduction of greenhouse gas emission. Human history shows that we humans are hardwired to take action when faced with an immediate threat. It was a necessity for the hunter/gatherers to defend themselves against wild animals. After the transition to agricultural societies, groups of humans had to fight off other groups from raiding their harvest in order to survive the winter. This necessity to protect against surprise attacks from other groups of humans has shaped the thinking and actions of most people to this very day. In contrast, climate change is a threat that endangers all the people on spaceship Earth. Therefore, it requires a comprehensive and coordinated joint action by all the nations.

In his book *Collapse: How Societies Choose to Fail or Succeed*, Jared Diamond [7] documents how societies, such as the Maya, the Anasazi of North America, and the Easter Islanders or the Vikings of Greenland, collapsed because of their abuse of the environment and their inability to change their mode of thinking and acting. He ends his book with the observation that while the Easter Islanders were busy deforesting the highlands of their overpopulated island, they had no way of knowing how other societies collapsed before them in order to learn from their mistakes. He argues that we have a big advantage over these past collapsed societies in that we live in an age of instant communication and have easy access to abundant knowledge about the causes of past collapses. In short, Diamond professes a deep faith in the ability of the world's national governments to correctly analyze the danger posed by the climate change phenomenon and to mobilize their citizens to take corrective action.

Unfortunately, there is little evidence in human history to justify this faith. As already mentioned, we humans are conditioned to react forcefully to immediate real or perceived threats, but we are very slow in voluntarily changing our thinking habits. There seems to be a "Generalized Newtonian Law" governing our thinking and actions. "Unless forced to change by a strong and immediate force we tend to keep thinking and doing the same things." Even natural scientists and engineers are reluctant to accept new theories or pay sufficient attention to the laws of nature unless they are forced to do so by overwhelming experimental evidence. A good example is Alfred Wegener's theory of plate tectonics which was accepted only many years after his tragic death in Greenland. Engineers are sometimes also tempted to ignore reality. After the Challenger space shuttle disaster, the Nobel laureate Richard Feynman served on the accident investigation commission. He admonished NASA to remember that "for a successful technology, reality must take precedence over public relations, for nature cannot be fooled" [8].

This then is the fundamental difficulty. Most thoughtful analysts of the climate change challenge agree that the prevailing strong belief in the economic growth mantra as a means of lifting the living standards of all the people on our planet has to switch to a sustainability mantra in the near future, as expressed, for example, by Dennis Bushnell, chief scientist at the NASA Langley Research Center [9]. However, in our view, such a major change in outlook will be adopted by the vast majority of the globe's population only if the effects of climate change become so visible and harmful that a new majority consensus develops in most countries to take decisive action.

The analysis therefore has to turn from a discussion of climate change to an examination of the *RATE* of climate change. It is one thing to read about relatively small and slow changes in remote areas of the world, say the receding ice cover in the Arctic or the disappearance of the coral reefs on the Australian East Cost. It is quite another thing to watch the intrusion of water into areas close to home which causes irrefutable economic damage. The question therefore is: How much flooding of Miami Beach, New York, San Francisco, etc. does it take to change the consensus opinion? Will it be necessary to wait until half of Holland is under water? In his book *The Flooded Earth: Our Future in a World Without Ice Caps*, Peter Ward [10] warns that sea level rise will be an unavoidable part of our future. He projects that the seas will rise 3 feet by 2050 and 9 feet by 2100 even if we stopped all carbon emissions today. He notes that the effects of 3 feet of sea level rise will be massive and 9 feet will be catastrophic.

Of course, there will be many other harmful effects which are described in many other papers and books. In our view, it suffices to take a look at the measured sea level rise over the past 100 years, as shown in Fig. 1.4, to realize that it is only a question of time until a Global Energy Apollo Program will be initiated with the support of most major countries.

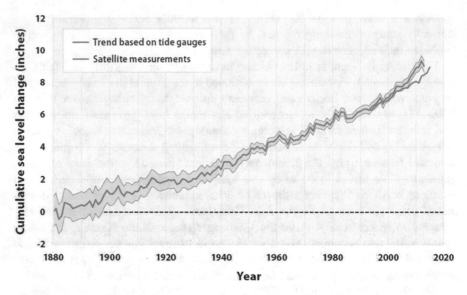

**Fig. 1.4** Measured sea level rise
Source: CSIRO (2015), NOAA (2016)

We argue that the start of such a program should not be delayed any longer because of another danger which has remained largely unrecognized in the general public, namely, the possibility of an ecological system shifting abruptly and irreversibly from one state to another when forced across critical thresholds [2]. Environmental scientists agree that global climate change is already causing major environmental effects, such as changes in the frequency and intensity of precipitation, droughts, heat waves and wildfires, rising sea level, water shortages in arid regions, new and larger pest outbreaks afflicting crops and forests, and expanding ranges for tropical pathogens that cause human illness.

However, there exists a well-justified fear that worse is in store because of the existence of "tipping points," i.e., thresholds beyond which a small additional increase in average temperature or some associated climate variable produces major changes to the whole system. Possible tipping points are the complete disappearance of Arctic sea ice in summer, leading to drastic changes in ocean circulation and climate patterns across the whole Northern Hemisphere; acceleration of ice loss from the Greenland and Antarctic ice sheets, driving rates of sea level increase to 6 feet or more per century; and ocean acidification from carbon dioxide absorption, causing massive disruption in ocean food webs. The existence of tipping points can be appreciated by anyone who is familiar with the effect of stalling an airplane. As soon as the pilot allows the wing to exceed the stall angle of attack by only a small amount, the flow pattern over the wing changes dramatically, causing the airplane to go into a steep dive. Many pilots and passengers have lost their lives because the pilots were unable to control their aircraft. Similarly, exceeding the climate tipping point will make it impossible to return to the preindustrial global climate even after the transition to an emission-free economy. Instead, the ice will continue to melt until its complete disappearance.

As already mentioned, it will become clear only in hindsight whether we have already passed the climate tipping point. The best estimate seems to point toward the mid-century. In any case, it will be prudent to stop carbon dioxide emissions as soon as possible. We humans find ourselves in a situation similar to the driver who plans to drive through Death Valley in mid-summer while knowing that the car radiator has a slight leak. Being short on money, he may gamble and take the risk. We are passengers in a spaceship whose air conditioning system shows signs of malfunctioning. We find ourselves in a race against time with the choice of repairing the system or risking a system crash.

These considerations lead us to the question whether it is possible to repair our spaceship before it is too late. Accepting the widely accepted estimate of mid-century for the tipping point, the challenge facing the global engineering community is to define the basic objective, analyze possible options to eliminate greenhouse gas emissions, and rank these options according to well-defined criteria. We are writing this book in the hope of being able to contribute to this objective.

# References

1. S. Arrhenius, On the influence of carbonic acid in the air upon the temperature on the ground. Phil. Mag. Fifth Ser. **41**(251) (1896)
2. A.D. Barnosky, E.A. Hadly, *Tipping Point for Planet Earth* (St. Martin's Press, New York, 2016)
3. W. Steffen et al., *Trajectories of the earth system in the anthropocene*. Proc. Nat. Acad. Sci. **9** (2018). https://doi.org/10.1073/pnas.1810141115
4. J. Hansen, *Storms of My Grandchildren* (Bloomsbury, New York, 2009)
5. A. Gore, *Our choice* (Melcher Media, New York, 2009)
6. N. Rich, *Losing Earth,* New York Times Magazine, 5 August 2018, pp. 1–70
7. J. Diamond, *Collapse* (Penguin Books, London, 2005)
8. R. Feynman, *What Do you Care What Other People Think?* (W.W. Norton, New York, 2001)
9. D. Bushnell, Where Is It All Going? Prospects for the Human Future, AAI Foresight Inc., www.aaiforesight.com (2016)
10. P.F. Ward, *The Flooded Earth* (Basic Books, New York, 2010)

# Chapter 2
# Current Status of Global Energy Consumption, Production, and Storage

As shown in Part II, the internationally accepted measuring unit of energy is 1 J, which is the energy exerted by the force of 1 N to move a mass of 1 kg a distance of 1 m. This definition of energy invokes the connotation with the physical labor of moving an object and is derived from Newton's second law. However, in later years, it was recognized that there are many forms of energy. Therefore, a better definition is to define energy as the ability to transform a system, a process that can involve any kind of energy. For example, a calorie is defined as the amount of heat needed to raise the temperature of 1 g of water from 14.5 °C to 15.5 °C. The experiments of the German physician Julius Mayer in 1842 and of the English physicist James Joule in 1847 led to conclusion that heat can be expressed in dynamic units so that 1 cal = 4.1855 J. These experiments led to the recognition that heat flowing into a system is equal to the increase of the system's internal energy and the work done by the system and thus to the formulation of the principle of the conservation of energy (i.e., the first law of thermodynamics).The German theoretical physicist Rudolf Clausius in 1850 then recognized that there can never be a positive heat flow from a colder to a hotter body which led him to the formulation of the second law of thermodynamics and the introduction of the term entropy in 1865. The second law states that the energy of the universe is fixed, but its distribution is uneven, and thus its conversions seek a uniform distribution causing the entropy of the universe to tend to a maximum. In practical terms it means that the conversion of highly ordered (low entropy) forms of energy to heat, a dispersed, low-quality, disordered (high entropy) form of energy, is irreversible. Heat can never be completely converted into another form of energy because only a portion of the initial input ends up as kinetic energy or as electricity. Most modern textbooks introduce the energy conservation principle as self-evident. However, Niels Bohr, the pioneer of quantum mechanics, briefly flirted with the idea that it is not conserved when he tried to explain some puzzling experiments in radioactivity. So did Lev Landau, another highly respected physicist who struggled with understanding the origin of the energy source of stars.

M. F. Platzer, N. Sarigul-Klijn, *The Green Energy Ship Concept*, SpringerBriefs in Applied Sciences and Technology, https://doi.org/10.1007/978-3-030-58244-9_2

By now, energy conservation is firmly established and accepted as a governing physical law.

Different forms of energy are often measured in different units. For example, electric energy is often measured in kilowatt-hour, heat energy in calories, and the chemical energy content of coal, oil, and natural gas in ton oil equivalent (t.o.e), but because of the energy conservation law, the measuring units for various forms of energy are directly related to each other. We adopt David MacKay's recommendation in his book *Renewable Energy – without the hot air* [1] to express all forms of energy in kilowatt-hour (kWh) because nobody except specialists can relate to "a barrel of oil" or "a million BTUs" (British thermal units), but most people pay their electricity bills for the kilowatt-hours used in a month. Therefore, most people can also relate to the power (i.e., the rate at which energy is used or produced by one person in 1 day). The kilowatt-hour per person per day (kWh/p/d) then provides a good illustration of the power produced or used in various forms (where it is understood that the words "produced" or "used" are merely sloppy words for indicating that the power is obtained from or converted to another form in compliance with the energy conservation law).

MacKay [1] lists the various needs and activities a person in a developed country takes for granted. They consist of being able to drive a car daily a distance of, say, 50 km; to take an intercontinental flight once per year; to heat and cool one's apartment or house; to provide proper lighting for one's home and office; to be able to buy and use various gadgets, such as a television, a computer, etc.; to be able to consume and produce food, potable water, etc.; and to be able to buy, transport, and use "stuff" (lumping together all additional items not used daily but acquired once in a while, such as a refrigerator, etc.). Finally, he also adds the need for public services, including the need for military and police forces. He estimates the daily power consumptions for all these needs as 40 kWh for car use, 30 kWh for jet flights, 37 kWh for heating and cooling, 4 kWh for lighting, 5 kWh for gadgets, 15 kWh for food and food production, 60 kWh for stuff, and 4 kWh for public services, arriving at a total daily power consumption per person of 195 kWh. He states that the average American consumes about 250 kWh per day, the average citizen of Bahrain and the United Arab Emirates exceeds 350 kWh, whereas citizens of lesser developed countries consume about 20 kWh which was the average power consumption before the start of the industrial revolution. The power consumption of the "average European" then can be assumed to be 125 kWh if one deletes the use of cars and airplanes.

Dividing this number by 24 in order to get the hourly power consumption per person per hour and multiplying it with 7 billion as the world population number in 2010 gives approximately 36.5 TW. On the other hand, the global primary consumption in 2010 was only 16.6 TW, clearly showing the need for a doubling of the global power production in order to raise the living standards of all people on our planet to the standard of the average European.

In the report GECO 2016 "Global Energy and Climate Outlook Road from Paris" by the European Commission's Joint Research Center [2], the world population is projected to grow to 8.5 billion in 2030 and to 9.75 billion in 2050, while the power

**Table 2.1**   Power densities of various sources

| Nuclear power generation density (reactor) | 50–300 MW/m$^2$ |
| --- | --- |
| Nuclear power generation density (land area for mining, processing, storage, etc.) | 230 W/m$^2$ |
| Coal mining power densities | 100 W/m$^2$–2 kW/m$^2$ in underground mines |
|  | 2 kW/m$^2$–33 kW/m$^2$ in open pit mines |
| Oil extraction power densities | 10–20 kW/m$^2$ |
| Natural gas extraction power densities | 10–15 kW/m$^2$ |
| Ethanol power production density (Brazil) | 0.45 W/m$^2$ |
| Ethanol power production density (USA) | 0.22 W/m$^2$ |
| Hydroelectric power production density (reservoirs) | 0.4–3 W/m$^2$ |
| Hydroelectric power generation density (river dams) | 1–500 W/m$^2$ |
| Wind power generation density (land) | 1.3–7 W/m$^2$ |
| Wind power generation (offshore) | 10–22 W/m$^2$ |
| PV power generation density | 3–30 W/m$^2$ |
| Concentrating solar power generation density | 25 W/m$^2$ |
| Tidal power generation density (Rance, Bay of Fundy) | 14 16 W/m$^2$ |

demand is expected to be 24 TW in 2030 and 29 TW in 2050. The share of total renewable power (consisting of conventional hydropower, biomass, wind and solar power, and other renewable power) is projected to increase from 13% in 2010 to 19%, mainly due to the increased contribution from wind and solar power.

The basic reason for this rather small projected increase in renewable power is found in the order of magnitude differences in power or energy densities between nuclear, fossil-based, and renewable densities shown in Table 2.1 [3].

Nuclear power production density far exceeds fossil-based power production density if only the land area required for the nuclear reactor is used as selection criterion. However, it decreases dramatically if the land claims for the complete fuel cycle (mining and processing of ores, uranium enrichment, and production of fuel elements, fuel reprocessing, and storage of radioactive wastes) are included. Renewable power production requires the largest land area because of the low power densities of all forms of renewable power generation, such as solar, wind, hydroelectric, tidal, wave, biomass, or geothermal power production. For this reason, fossil-based and nuclear power production have been the favored production methods. A second selection criterion is the economic viability of the power production method. These two criteria played the major role in the past together with the requirement that an adequate supply of power must be available at any moment. However, in recent decades the environmental impact of different power production methods has been recognized as a third criterion which must be incorporated into the decision-making process. Barros et al. [4] developed a methodology which allowed the assessment of the sustainability of five conventional and five renewable power production methods. We agree with this methodology and therefore limit the discussion to the four most promising renewable power production methods, namely, photovoltaic, concentrating solar, hydroelectric, and wind power production.

Photovoltaic (PV) panels convert sunlight directly into electricity by means of the photovoltaic effect. The conversion efficiency varies between 10% and at most 20% for very expensive panels. The theoretical conversion limit (Shockley-Queisser limit) is 31%. The solar power density varies between 80 and 280 $W/m^2$ depending on the geographical location. According to MacKay [1], a typical value for England is 110 $W/m^2$. The average power from a south-facing panel therefore is given by 20% $\times$ 110 $W/m^2$ = 22 $W/m^2$. Allotting to each person approximately 10 $m^2$ and assuming 12 h of sunshine will deliver approximately 2.5 kWh per person per day. Since the average energy demand in Europe is 125 kWh/p/d, an area of 3000 $m^2$ would be required to satisfy this energy demand. These estimates show the difficulty of providing adequate amounts of energy produced by solar panels alone.

For this reason, it is tempting to use mirrors or lenses to concentrate solar radiation in order to heat water which drives a steam turbine connected to an electric power generator. Heat storage in molten salts allows to continue generating after sunset. This technology is still in the development stage. In 2009 an organization called DESERTEC Industrial Initiative, comprising Siemens, Munich Re, E.ON, Deutsche Bank, and others, proposed the construction of wind farms and concentrating solar power plants in sunny Southern European, Middle Eastern, and North African countries and to deliver the electric power to the remainder of Europe via high-voltage direct-current (HVDC) transmission lines. This type of transmission technology has been shown to transmit electricity over large distances at smaller losses than conventional high-voltage alternating-current lines and to require less physical hardware and land area. The ultimate objective of the DESERTEC project was to provide the one billion people of Europe and adjacent countries with 125 kWh/p/d, i.e., to satisfy Europe's total electrical power demand. The average power per unit land area delivered by concentrating solar power plants is estimated to be 15 $W/m^2$ which translates into a square of 600 by 600 km or the total land area of Germany. By 2014 various technical, economic, and political complicating factors convinced the sponsoring companies to abandon these ambitious plans. As of 2017 concentrating solar power plants delivered less than 2% of worldwide installed capacity of solar electricity plants.

In contrast, hydroelectric power production began on a small scale already in 1882 and was rapidly applied in the Alpine countries, Scandinavia, and the United States. According to Smil [3], the reservoirs behind the world's large dams cover an area nearly twice as large as Italy, and the world's largest hydroelectric power plants yield about 3 $W/m^2$, whereas average power plants yield only about 1.5 $W/m^2$, showing again the difficulty of being able to satisfy the global power demand. In fact, the potential of building additional large plants is quite limited.

Modern wind turbine developments started in the 1970s with the construction of turbines having a rotor diameter of only 17 m and a power output of 75 kW and progressed to 125 m with a power output of 5 MW. This year General Electric is offering a turbine with a diameter of 250 m and an output of 12 MW. Windy sites with annual mean wind speeds of 7–7.5 m/s produce power densities of 400–500 W/m$^2$. The conversion of the kinetic wind energy in a given wind stream is limited by Betz' law which states that at most 16/27 (59%) can be captured. Actual power conversion efficiencies are between 30 and 40%. Furthermore, the seasonal wind variations typically allow the annual capture of only 30–35% of the rated turbine output (the capacity factor). As a result, the most densely packed wind farms reach a power density up to 15 W/m$^2$, and more typical sites have a density of 5–7 W/m$^2$. Offshore sites have larger average wind speeds, and therefore offshore wind turbines reach densities between 10 and 22 W/m$^2$.

Among these four most promising renewable power generation methods, only the hydroelectric plant can deliver power at a steady rate, whereas the other plants are subject to large variations in the energy inputs due to the daily or seasonal variations in solar and wind conditions. Hence the selection of an adequate energy storage method becomes an equally important criterion. Current energy storage technologies comprise hydroelectric pumped storage, compressed air energy storage, battery energy storage, flywheel energy storage, warm water storage, phase change materials storage, power-to-gas/liquid (hydrogen, methane, methanol, etc.) energy storage, thermochemical energy storage, fossil fuels (coal, oil, natural gas), biomass, and biodiesel [5]. Pumped-storage hydroelectric power stations are the only available large-scale storage technology for electricity. They represent the most mature storage technology, but their development potential is constrained by geographical requirements. A considerable amount of experience also exists for compressed air energy storage. Flywheel storage has so far been used only in industrial applications. They provide high power output but have only a small capacity. The storage of electrical energy in batteries is a well-established technology, but the battery capacity is limited by the mass of the electrode materials and the electrical power by the rates of the reactions. High power density therefore requires large porous electrode surfaces. It appears that the most promising technology for the storage of large quantities of energy over extended periods of time is the storage of hydrogen produced from water electrolysis. Three electrolytic processes are suitable to split water into hydrogen and oxygen, namely, alkaline, proton-exchange membrane (PEM), and high-temperature electrolysis. Alkaline electrolysis is already widely used in the chemical industry, PEM electrolysis is a relatively young technology and currently requires precious metals such as platinum and iridium as catalysts, and high-temperature electrolysis has so far been used only on a laboratory scale for research purposes. Small quantities of hydrogen can be stored in spherical or cylindrical storage tanks. Examples are the pressurized hydrogen gas tanks used in commercially available hydrogen fuel cell cars. Large quantities are mostly stored in caverns, such as the Clemens Dome and Moss Bluff Salt Dome caverns in Texas. Small quantities of hydrogen can be transported in tankers as pressurized gas or liquefied at −253°C. Large quantities can be transported in pipelines. For example, several European companies have already successfully operated hydrogen gas pipelines for decades. Hydrogen can also be transported in natural gas pipelines as long as the hydrogen content does not exceed 5%.

An additional option to store large amounts of energy is to convert hydrogen into other forms of energy by reacting the hydrogen with carbon dioxide to produce methane, methanol, higher fuels, or chemicals. Conversion to methane makes it possible to use the existing natural gas grid, and conversion to methanol or to Fischer-Tropsch fuels makes possible the transportation by road, rail, and pipeline. All these fuels remain climate neutral if the hydrogen conversions are carried out with carbon dioxide extracted from the air, water, or biogenic sources. Hydrogen can also be converted to ammonia by means of the Haber-Bosch process. Although the main use of ammonia today lies in the production of fertilizers, the use of ammonia as an energy storage device may become quite important in the future because there already exists an extensive ammonia pipeline and tank infrastructure apart from the fact that the existing natural gas or petroleum pipelines could be easily converted to carry ammonia. Like hydrogen, ammonia can burn directly in spark-ignited internal combustion engines or may be used to produce electricity in solid oxide, proton-conducting ceramic, and molten salt direct ammonia fuel cells with only nitrogen and water vapor as combustion products. Like hydrogen, ammonia is lighter than air and is not a greenhouse gas. Ausfelder et al. [5] provided a very comprehensive review of various energy storage systems. Leighty and Holbrook [6] discussed the potential of converting hydrogen into ammonia and transporting it through existing pipelines. Gahleitner [7] also reviewed the various hydrogen storage possibilities and provided a comprehensive list of the power-to-gas pilot plants and of the hydrogen production and utilization and periods of operation of these plants.

# References

1. D.J.C. MacKay, *Sustainable Energy –Without the Hot Air* (UIT Cambridge, Cambridge, 2009)
2. GECO 2016 Global Energy and Climate Outlook, Road from Paris, European Commission Joint Research Center Science for Policy Report, July 2016
3. V. Smil, *Energy in Nature and Society* (MIT Press, Cambridge, MA, 2008)
4. C.J.J. Barros, M.L. Coira, M.P.C. Lopez, Assessing the global sustainability of different electricity generation systems. Energy **89**, 473–489 (2015)
5. F. Ausfelder et al., Energy storage as part of a secure energy supply. ChemBioEng. Rev. **4**(3), 144–210 (2017)
6. W.C. Leighty, J.H. Holbrook, Running the world on renewables: Alternatives for transmission and low-cost firming storage of stranded renewables as hydrogen and ammonia fuels via underground pipelines, ASME International Mechanical Engineering Congress, Houston, Texas, IMECE2012–87097, 9–15 Nov 2012
7. G. Gahleitner, Hydrogen from renewable electricity: An international review of power-to-gas pilot plants for stationary applications. Int. J. Hydrog. Energy **38**, 2039–2061 (2013)

# Chapter 3
# Climate Tipping Points and Climate Irreversibility

As shown in Fig. 3.1a and 3.1b, the global greenhouse gas emissions are expected to increase from 48.3 GtCO2e (gigatons of equivalent CO2) in 2010 to 63.1 GtCO2e in 2030 and to 73.6 GtCO2e in 2050. This amounts to a yearly emission of 6.5 tCO2e per person. The total amount of carbon dioxide emitted into the atmosphere from 1870 to 2014 is estimated to be 2000 GtCO2, which corresponds to 545 Gt of carbon because the ratio of the atomic weights of $CO_2$ and carbon is 44/12. In their paper "Assessing Dangerous Climate Change," Hansen and co-authors [1] warn that cumulative carbon emissions of 1000 GtC will cause an eventual warming of 3–4 °C with disastrous consequences. At the Paris Climate Conference, the participating countries pledged to reach a reduction of the total energy demand by 9% by 2050, as a result of fuel substitution and energy efficiency. This is expected to reduce the greenhouse gas emissions to 50 GtCO2e in 2050, hence to decline to the 2010 emission level (Fig. 3.1b). This reduction is clearly insufficient to "bend the curve" to the point of zero $CO_2$ emission by 2050 in order to prevent irreversible and possibly quite abrupt climate change.

Although there are still many people who deny the phenomenon of human-induced climate change, there are many others who tend to visualize climate change as a gradual and smooth increase in scale and severity of impacts with increasing temperature. This leads to the image that the impact of, say, a global temperature rise slightly in excess of 2 °C is somewhat similar to 2 °C, only a bit worse. This assumption of a smooth relationship between temperature and level of impact leads to the notion that there is an optimal temperature which allows the balancing of climate mitigation costs versus climate impact and adaptation costs. Unfortunately, it is now becoming reasonably well understood that, as the planet warms, it may be approaching several thresholds that may result in significant step changes in the level of impacts once triggered.

These thresholds are now usually denoted as "tipping elements" and "tipping points." A tipping point is defined as a critical point where a relatively small change in climate forcing can commit a system to a qualitative change in state. Hence the

M. F. Platzer, N. Sarigul-Klijn, *The Green Energy Ship Concept*, SpringerBriefs in Applied Sciences and Technology, https://doi.org/10.1007/978-3-030-58244-9_3

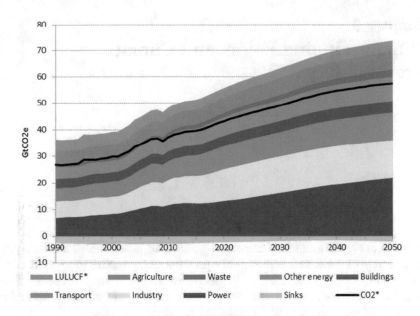

**Fig. 3.1a**   Past and projected greenhouse gas emissions
Source: GECO 2016 *Global Energy and Climate Outlook, Road from Paris,* European Commission
Joint Research Center Science for Policy Report, July 2016 LULUCF (Land Use, Land Use
Change, Forestry)

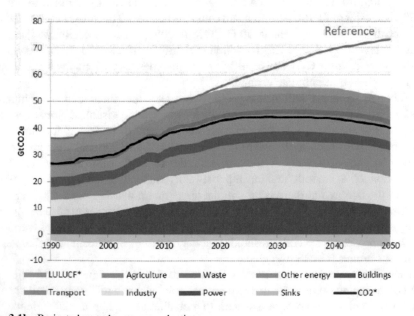

**Fig. 3.1b**   Projected greenhouse gas reductions
Source: GECO 2016 *Global Energy and Climate Outlook, Road from Paris,* European Commission
Joint Research Center Science for Policy Report, July 2016

word "tipping point" is introduced to convey the notion that "a small change can make a big difference." The term "tipping element" has been introduced to describe those large-scale components of the Earth system that could be forced past a "tipping point" to undergo a transition to a quite different state.

According to [2], tipping elements should satisfy the following conditions:

(a) Be components of the Earth system that are at least subcontinental in scale.
(b) The factors affecting the system can be combined into a single control.
(c) There exists a critical value of this control (the tipping point) from which a small perturbation leads to a qualitative change in a crucial feature of the system, after some observation time.

Although deliberately broad and inclusive, this definition includes cases where the transition is faster than the forcing causing it. It is then called "abrupt" or "rapid" climate change. On the other hand, in some cases passing the tipping point is barely perceptible, yet leads to large changes in the future. Furthermore, passing the tipping point can include transitions that are reversible (where reversing the forcing will cause recovery at the same point it caused collapse) and those that exhibit irreversibility (within humanly meaningful time horizons).

The following tipping elements have been identified:

(a) Tipping elements associated with warming and the melting of ice at high latitudes or altitudes, namely, systems which are in the Arctic region and comprise sea-ice and the Greenland ice sheet and systems whose tipping point is probably more distant and uncertain, namely, the West Antarctic sheet, continental ice caps, permafrost, and the boreal forests
(b) Tipping elements that can influence other tipping elements around the world, namely, the Atlantic thermohaline circulation and the El Nino Southern oscillation
(c) Tipping elements associated with changes in hydrological systems in the tropics and expanding subtropics, namely, the Amazon rainforest, the West African Monsoon and Sahel, the Southeast Asian monsoons including the Indian Summer Monsoon, and Southwest North America

An additional detailed discussion of the danger of tipping points was recently published by Barnosky and Hadly [3].

The inclusion of tipping elements leads to an estimate of a 0.5 m sea level rise by 2050 versus 0.15 m without tipping elements. In addition to the rise in sea level, increased sudden-onset coastal flooding events impact people and assets. 90% of the exposed assets are in eight countries (China, the USA, India, Japan, Netherlands, Thailand, Vietnam, Bangladesh). Detailed regional impact estimates due to projected sea level associated with tipping elements, impacts of monsoon interference on India, and hydrological impacts on Amazonia and of increased aridity in Southwest North America are given in [2]. These estimates force the insurance companies to account for them in setting insurance rates. A similar impact assessment for the United States was conducted under the leadership of Michael Bloomberg, former mayor of New York; Henry Paulson, Jr., former secretary of the US Treasury; and Thomas Steyer, founder of Farallon Capital Management LLC. They project that by 2050 between \$66 billion and \$106 billion worth of existing coastal property will be below sea level nationwide, with \$238 billion to \$507 billion worth of property below sea level by 2100. By 2050 the average American will likely see 27–50 days over 90° Fahrenheit each year – two to more than three times the average annual number of 95° days over the past 30 years. During portions of the year, extreme heat will surpass the threshold at which the human body can maintain a normal core temperature without air conditioning. Therefore, demand for electricity for air conditioning will surge in the hot sections of the United States. The detailed impacts for the various regions in the United States are described in [4].

# References

1. J. Hansen et al., Assessing dangerous climate change: required reduction of carbon emissions to protect young people, future generations and nature. PLOS one (2013). https://doi.org/10.1371/journal.pone.0081648
2. World Wide Fund for Nature (formerly World Wildlife Fund), *Major Tipping Points in the Earth's Climate System and Consequences for the Insurance Sector* (WWF, Gland, 2009)
3. A.D. Barnosky, E.A. Hadly, *Tipping Point for Planet Earth* (Thomas Dunne Books/St. Martin's Press, New York, 2016)
4. Risky Business: The Economic Risks of Climate Change in the United States. A Climate Risk Assessment for the United States, June 2014

# Chapter 4
# Review of Past Energy Transitions

Recently, Sovacool [1] investigated the issue of the time scale of energy transitions. Based on the available empirical data, he presents a "mainstream" view of energy transitions as long, protracted affairs, often taking decades to centuries to occur. In this context, it is important to define the term "transition." Adopting the transition time as the time between the introduction of a new fuel and its rise to 25% of a national or global market share, the phase-in of coal in England took 160 years, in Germany 102, in France 107, and similar times in the other European countries. The shift again from coal to oil and electricity was more rapid, but it still ranged from 47 to 69 years. In the United States, crude oil took half a century from its exploratory stages in the 1860s to capturing 10% of the national market in the 1910s, then 30 more years to reach 25%. Nuclear electricity took 38 years to reach a 20% share in the United States which occurred in 1995. Globally, coal reached the 25% mark after more than 500 years of coal mining in 1871, and crude oil took some 90 years after its first discovery in Pennsylvania in 1859. Hydroelectricity, natural gas, wind power, and solar panels still have to surpass the 25% threshold. In the face of this and other evidence, Smil [2, p. 150] deduces that:

> Energy transitions have been, and will continue to be, inherently prolonged affairs, particularly in large nations whose high levels of per capita energy use and whose massive and expensive infrastructures make it impossible to greatly accelerate their progress even if we were to resort to some highly effective interventions.

However, there is contrary evidence that under certain conditions energy transitions can occur rather quickly [1]. Sweden accomplished an almost complete shift to energy-efficient lighting in about 9 years. China succeeded in switching to 185 million energy-efficient cook stoves between 1983 and 1998. Indonesia also implemented a large household energy saving program by converting from kerosene stoves to liquefied petroleum gas stoves, increasing the number of such stoves from a mere three million to 43.3 million in just 3 years from 2007 to 2009. Brazil substituted ethanol for petroleum in conventional vehicles between 1975 and 1981

M. F. Platzer, N. Sarigul-Klijn, *The Green Energy Ship Concept*, SpringerBriefs in
Applied Sciences and Technology, https://doi.org/10.1007/978-3-030-58244-9_4

so that 90% of all new vehicles sold in Brazil could run on ethanol. In the United States, air-conditioning units started to be mass produced in 1947 and 43,000 were sold that year. Six years later this number had increased to one million and in 1960 6.5 million were installed. In 1970 this number rose to 24.2 million, and by 2009 87% of single-family homes had air-conditioning. All American cars nowadays are air-conditioned so that the United States currently uses more energy for air-conditioning (about 185 billion kWh) than all other countries combined.

Further evidence for rapid energy supply transitions is found in Kuwait where it shifted to oil in about 9 years by 1950, and the Netherlands transitioned from oil and coal to natural gas in a little over a decade. After the 1974 oil crisis, France decided to switch its coal-, oil-, and gas-based electricity production to nuclear power plants. It built 56 reactors from 1974 to 1989, and, as a result, nuclear power grew from 4% of national electricity supply in 1970 to almost 40% by 1982. Denmark transitioned to wind energy in the 1970s and 1980s for electricity production and heating, making it the first major industrialized country (besides Iceland) to be almost completely powered by renewable energy sources (with additional hydropower imported from Norway). Another example is Ontario which started to retire all coal-fired electricity generation in 2003 and by 2014 had replaced it with wind, hydro, solar, and nuclear power. This switch required the investment of more than $21 billion, but is estimated to save 4.4 billion per year in health, environmental, and financial damages.

Admittedly, these examples of fast energy transitions mostly involve specific energy sectors or small countries. Nevertheless, they show that future transitions may occur equally fast when driven by social or political movements inspired by climate change and innovation. It is this aspect which we want to explore in more detail.

# References

1. B.K. Sovacool, How long will it take? Conceptualizing the temporal dynamics of energy transitions. Energy Res. Soc. Sci. **13**, 202–215 (2016)
2. V. Smil, *Energy Transitions: History, Requirements, Prospects* (Prager, Santa Barbara, 2010), p. 150

# Chapter 5
# Lessons from Past Major Engineering Initiatives

In a report published by the National Academy of Sciences (Technical Bulletin No.2, January 1941), Theodore von Karman, C.F. Kettering, Lionel S. Mark, and R.A. Millikan stated that "In its present state, and even considering the improvements possible when adopting the higher temperatures proposed for the immediate future, the gas turbine could hardly be considered a feasible application to airplanes mainly because of the difficulty in complying with the stringent weight requirements imposed by aeronautics. The present internal-combustion-engine equipment used in airplanes weighs about 1.1 pounds per horsepower, and to approach such a figure with a gas turbine seems beyond the realm of possibility with existing materials. The minimum weight for gas turbines even when taking advantage of higher temperatures appears to be approximately 13 to 15 pounds per horsepower."

This assessment of the development potential of jet propulsion was given to the United States Navy one and a half years after the first successful flight of the HE-178 aircraft on 27 August 1939. It was propelled by a single turbojet engine invented and developed by Hans von Ohain at the Heinkel Aircraft Company in Germany. It developed 838 pounds of thrust, generated by compressing the inlet airflow by a single-stage axial compressor and then by a centrifugal compressor, heating the air in a combustor and expanding the gas first through a turbine and then through an exhaust nozzle. Yet, in the United States, still being at peace in early 1941, the leading American aeronautical scientists greatly underestimated the need for superior speeds in air combat in spite of the disadvantages stated by the authors of Technical Bulletin No. 2. As a matter of fact, the representatives of the German Air Ministry were also unimpressed by Heinkel's demonstration flight, but they did support the development of a competitor aircraft at the Messerschmitt aircraft company, the ME-262, and of a competitor engine at the Junkers Aircraft Company, the JUMO 004. This engine was the first successful axial compressor turbojet engine, first run in 1940, and the ME-262 made its maiden flight in July 1942. As

Germany's war fortunes turned in 1942 and Allied bombing raids into Germany increased, the Air Ministry ordered the ME-262 into full production. It reached German air force units in September 1944, and by May 1945 some 1400 ME-262s had been produced. The superior performance characteristics of the ME-262 aircraft made it not only obvious that military aircraft had to transition to jet propulsion, but it also stimulated design studies of civilian transport airplanes. The de Havilland aircraft company began such design studies of a jet-propelled passenger airplane before the end of 1943 which eventually resulted in the de Havilland Comet 1 airplane. It was powered by four 4450 pound de Havilland Ghost centrifugal turbojet engines and could carry 36 passengers. It made its first flight on 27 July 1949, and nine planes started operation for British Overseas Airlines BOAC in May 1952. The Comet was able to halve the flying time of piston-engine-powered airplanes, but it impressed also by the smoothness of jet travel at "over the weather" altitudes in pressurized cabins. As a consequence, in spite of three fatal Comet accidents, the advantages of jet travel were quickly recognized in the Western world and in the USSR, inspiring the Boeing aircraft company to deliver the Boeing 707 and the Tupolev aircraft company the Tu-104 aircraft by the late 1950s, soon followed by the Douglas, Convair, and Lockheed aircraft companies. Hence it took only 20 years to revolutionize intercontinental air travel.

It is also important to remember that an equally rapid development occurred after the demonstration of the feasibility of powered flight by the Wright brothers in December 1903. Military needs were again the major stimulus so that by the end of World War I, well over 200,000 planes had been produced by the major combatant nations. The resulting surplus of planes and flight expertise quickly led to the establishment of commercial airline companies in the early 1920s.

Another revolutionary development started in 1942 when the German V-2 super-sonic rocket was successfully test fired and quickly ordered into series production. By war's end in May 1945, over 6000 missiles had been produced, and some 4320 had been fired between 6 September 1944 and 27 March 1945 against London and targets on the continent. The further development of this missile in the Soviet Union and the United States led to the launch of the first satellite "Sputnik" on 4 October 1957, soon followed by the first American satellite in January 1958. Hence it took less than 20 years to initiate an entirely new technology. The continuing Cold War tensions between the United States and the Soviet Union then led to the development of the human space flight technology in Projects Mercury, Gemini, and Apollo which culminated in the landing of two American astronauts in July 1969 only 12 years after "Sputnik." At its peak Project Apollo involved 530,000 people and cost $107 billion (in 2016 dollars).

Much more information about these pioneering developments can be found in references [1–5].

# References

1. P. Scott, *The Pioneers of Flight* (Princeton University Press, Princeton, 1999)
2. M.J. Neufeld, *Von Braun* (Alfred A Knopf, New York, 2007)
3. J. Harford, *Korolev* (Wiley, New York, 1997)
4. M. Conner, *Hans von Ohain* (American Institute of Aeronautics & Astronautics, Reston, 2001)
5. S.H. Verhovek, *Jet Age, the Comet, the 707, and the Race to Shrink the World* (Penguin Group, New York, 2010)

# Chapter 6
# Recent Analyses and Current Proposals for Sustainable Global Power Production

In the years 2008 and 2009, Hansen [1.4], Gore [1.5], and MacKay [2.1] expressed the following principal conclusions:

> Hansen, pages 204 and 205: "Yes, I know there is no silver bullet solution for world energy requirements. We need an urgent, substantial research and development program on fourth-generation nuclear power, so that we have at least one viable option in the likely event that efficiency and renewables cannot provide all needed energy."
>
> Gore, page 405: "The choice is awesome and potentially eternal. It is in the hands of the present generation."
>
> MacKay, page 112: "To sustain Britain's lifestyle on its renewables alone would be very difficult. A renewable-based energy solution will necessarily be large and intrusive."

However, Jacobson and Delucchi [1] argued in 2009 that a complete transition to emission-free power production can indeed be accomplished within 20 years using wind, water, and solar power only. They therefore call it the WWS program. Jacobson et al. followed up on their original WWS proposal [1] with the development of detailed information about the feasibility of providing the whole world, the United States, or California [2–5] with power derived exclusively from land or offshore-based wind, solar, and water resources. For example, they postulate that the global power demand satisfied by conventional fossil fuels and wood will increase from 12.47 TW in 2010 to 21.6 TW in 2050. Similarly, it will rise for the United States from 2.37 TW in 2010 to 3.08 TW in 2050 and for California from 206 GW in 2010 to 280 GW in 2050. Converting the California power demand to WWS power generation (by means of 25,211 additional 5 MW onshore wind turbines, 7809 offshore 5 MW turbines, 4963 wave devices, 72 additional geothermal plants, 3371 tidal turbines, 14,990,000 residential roof PV systems, 533,700 commercial/government PV systems, 3450 utility PV plants, and 1226 concentrating solar power plants), the total nominal power output of the existing and additional WWS power generators will be 624.4 GW. A total of 2.77% of California land area will be required for these power generators, although most of this area will still be usable for multiple purposes, such as grazing, agricultural purposes, etc. In [5] they

M. F. Platzer, N. Sarigul-Klijn, *The Green Energy Ship Concept*, SpringerBriefs in
Applied Sciences and Technology, https://doi.org/10.1007/978-3-030-58244-9_6

address the concerns of utilities and grid operators that WWS systems cannot accommodate their inherent power supply variability problems caused by the daily and seasonal wind and solar power fluctuations. To this end they coupled numerical simulations of time- and space-dependent weather with simulations of time-dependent power demand, storage, and demand response. They obtained solutions by storing electricity in hydrogen tanks or in phase-change materials or using pumped hydropower systems as well storing heat in soil and water and cold in water and ice. Clack et al. [6] questioned the feasibility of providing low-cost solutions to the grid reliability problem with the WWS approach on the grounds that there are no electric storage systems available today that can affordably and reliably store the vast amounts of energy needed over weeks to reliably meet demand using expanded wind and solar power alone. Jacobson et al. responded to this criticism in a detailed analysis in reference [5].

In 2016, Brick and Thernstrom [7] examined the question whether a combination of low-carbon base load generation technologies (i.e., nuclear and fossil fuels with carbon capture and sequestration), along with a 25% contribution from wind and solar, can achieve the same emission reductions at less cost than a WWS system. They concluded that a WWS system must be significantly larger than a combined system because intermittent resources like wind and PV have low capacity factors, and therefore a significantly larger system with additional power storage and transmission capability is required to reliably generate the same output power. As a result, the system cost and capital requirements are substantially increased together with an increase in land use. They therefore conclude that the transition to a 100% renewable power generation system implies the commitment to building the largest possible power system. They also emphasize that any new energy infrastructure – whether nuclear power plants, hydroelectric dams, transmission lines, or industrial-scale wind farms and PV installations – inevitably generates never-ending political and legal battles.

In the same year, David King, John Browne, Richard Layard, Gus O'Donnell, Martin Rees, Nicholas Stern, and Adair Turner of the United Kingdom [8] proposed a Global Apollo Program with the objective to start a 10-year RD&D program with the following features:

(1) Target: to produce new-build base-load energy from renewable sources cheaper than new-build coal in sunny parts of the world by 2020 and worldwide from 2025.
(2) Scale: pledges from governments joining the consortium to spend an annual average of 0.02% of GDP as public expenditure on the Program from 2016 to 2025.
(3) Roadmap: to generate year by year a clear roadmap of the scientific breakthroughs required at each stage to maintain the pace of cost reduction. A Roadmap Committee of some 20 senior technologists and businessmen will construct and revise the roadmap year by year.

Also in the same year, F. P. Incropera [9] concluded his discussion of the climate change problem with the following statement: "Reducing emissions will require a sharp departure from the current norm of associating living standards with levels of consumption. It will also require all hands on deck. As individuals, organizations, and governments, the entire global community must be part of the solution. Will this story have a happy ending? Will the effects of global warming be manageable? At this time, there are no answers to these questions. But what can be said is that, through its action or inaction, humankind will determine its own destiny."

# References

1. M. Jacobson, M. Delucchi, A path to sustainable energy by 2030. Sci. Am. **301**, 58–65 (2009)
2. M.A. Delucchi, M.A. Jacobson, Providing all global energy with wind, water, and solar power, part II: reliability, system and transmission costs, and policies. Energy Policy **39**(3), 1170–1190 (2011)
3. M.Z. Jacobson, M.A. Delucchi, M.A. Cameron, B.A. Frew, Low cost solution to the grid reliability problem with 100% penetration of intermittent wind, water, and solar for all purposes. Proc. Natl. Acad. Sci., PNAS Early Edition, 1–6 (2016)
4. M.Z. Jacobson et al., A roadmap for repowering California for all purposes with wind, water, and sunlight. J. Energy **73**, 875–889 (2014)
5. M.Z. Jacobson, M.A. Delucchi, M.A. Cameron, B.V. Mathiesen, Matching demand with supply at low cost in 139 countries among 20 world regions with 100% intermittent wind, water, and sunlight (WWS) for all purposes. Renew. Energy. Accepted 2 Feb 2018. https://doi.org/10.1016/j.renene.2018.02.009
6. C.T.M Clack et al., Evaluation of a proposal for reliable low-cost grid power with 100R% wind, water, and solar. Proc. Natl. Acad. Sci., Early Edition, approved 24 Feb 2017
7. S. Brick, S. Thernstrom, Renewables and decarbonization: studies of California, Wisconsin and Germany. Electr. J. **29**, 6–12 (2016)
8. D. King, J. Browne, R. Layard, G. O'Donnell, M. Rees, N. Stern, A. Turner, *A Global Apollo Programme to Combat Climate Change* (Center for Economic Performance, The London School of Economics and Political Science, 2015)
9. F.P. Incropera, *Climate Change: A Wicked Problem* (Cambridge University Press, Cambridge, 2016)

# Chapter 7
# Problem Definition

Given these assessments let us clearly define the challenge. The carbon dioxide emissions need to be reduced to zero by 2050 in order to prevent irreversible and possibly quite abrupt climate change. The challenge confronting the global engineering community therefore is to show that there are options to reach this goal by mid-century. The analysis has to be conducted by adhering to well-established procedures and criteria. In our view, the task which needs to be accomplished can be stated quite succinctly as follows:

Satisfy the GEA 2016 specified global power demand of 24 TW in 2025 and of 29 TW in 2050 by emission-free power production

Possible approaches to satisfy this global demand by renewable power production methods need to be evaluated according to the following criteria:

(a) Ranking by renewable energy availability
(b) Ranking by power density
(c) Ranking by land use
(d) Ranking by capacity factor
(e) Ranking by economic viability
(f) Ranking by energy storability
(g) Ranking by socio-political acceptability

It is apparent from the review of previous proposals that the wind-water-solar (WWS) approach presented by Mark Jacobson and associates is the only sufficiently specific proposal to allow a quantitative evaluation and comparison with our own "energy ship" or "wind over water" (WoW) proposal presented below. We therefore first present the major features of this proposal and compare it with the WWS approach by ranking it according to the above criteria.

M. F. Platzer, N. Sarigul-Klijn, *The Green Energy Ship Concept*, SpringerBriefs in Applied Sciences and Technology, https://doi.org/10.1007/978-3-030-58244-9_7

# Chapter 8
# The Energy Ship Concept

This concept of power generation is based on the recognition that most of the global wind power is to be found over the oceans. Therefore, it is logical to ask the question whether it is possible to exploit this vast wind energy resource. As shown in Fig. 8.1, there are vast ocean regions with persistently high wind speeds which offer significantly higher power densities and capacity factors than available for onshore and offshore wind turbines. The power densities range between 500 and 1000 W/m$^2$, and the capacity factors are likely to be twice the offshore turbine factors. The ocean wind power available in these regions is conservatively estimated as 250 TW, hence far in excess of the projected global power demand of 29 TW by mid-century.

In recognition of this fact, we proposed in 2009 [1] a renewable power production system which avoids a lengthy approval process because it moves the power production process into remote ocean areas. We suggested that it is technically feasible to build sailing ships of a size comparable to modern container ships which can produce about 2 or 3 MW of electric power for electrolytic conversion of seawater into hydrogen and oxygen. A major advantage of this concept is the conversion of wind power into storable energy in the form of high-pressure or liquefied hydrogen. Another advantage is the much higher capacity factor available to sailing ships operating in wind-rich ocean areas as compared to land- or offshore-based wind turbines. These two features make it possible to implement the "power-to-gas/liquid" concept, whereby the intermittency problem of conventional wind and solar power plants can be overcome by making hydrogen available on demand. A further advantage is the ability to operate the energy ships in international ocean areas so that any country can send its fleet of energy ships to these areas without having to ask for permission. Therefore, the legal obstacles encountered by renewable energy development projects within a country's national boundaries are minimized.

Salomon seems to have been first in 1982 to recognize the feasibility of this method for large-scale energy conversion [2], followed by the additional patent

M. F. Platzer, N. Sarigul-Klijn, *The Green Energy Ship Concept*, SpringerBriefs in
Applied Sciences and Technology, https://doi.org/10.1007/978-3-030-58244-9_8

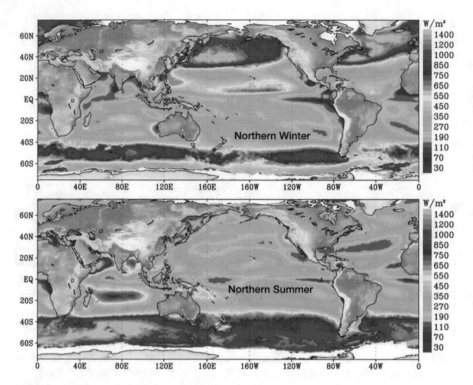

**Fig. 8.1** Global ocean wind power distribution
Source: NASA Earth Observatory

disclosures [3–5]. Kim and Park proposed to exploit the winds at altitude by propelling the ship with huge parawings [6].

The basic building blocks of the "energy ship concept" are shown in Fig. 8.2. A sailing ship cruises at the optimum apparent wind angle which roughly corresponds to the beam reach sailing mode (see Chap. 22). The ship is equipped with a hydrokinetic turbine + electric generator whose electric power output is either stored in electric batteries or is fed into a desalinator and electrolyzer for the purpose of splitting the seawater into hydrogen and oxygen. The hydrogen is either compressed and stored in suitable tanks at pressures of 350–700 bar or liquefied and stored in suitable liquefied hydrogen containers. The oxygen is either released into the atmosphere or stored in suitable oxygen tanks. The hydrogen (and oxygen) tanks then are periodically transferred to special tankers for transportation to coastal ports. The hydrogen can be used directly for cooking and heating purposes, or it is reconverted into electricity in special hydrogen power plants or fuel cells. This reconversion process yields potable water as a welcome secondary product.

It is worth noting that the energy ship concept makes it possible to store large amounts of energy in the form of compressed or liquefied hydrogen without recourse to new technology. It merely requires the application of the well-established sailing

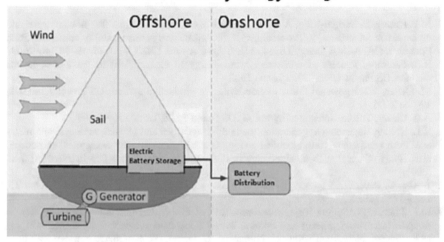

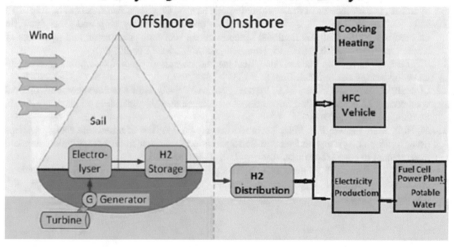

**Fig. 8.2** The energy ship concept [9]. (**a**) Electric battery energy storage; (**b**) hydrogen production and delivery

ship technology in combination with hydrokinetic turbine, desalination, and electrolysis technology as well as hydrogen compression or liquefaction and storage technology. Each one of these technologies is described in more detail in the second part of the book and in references [7–14]. The major elements and their developmental status are summarized in the next chapter.

# References

1. M.F. Platzer, N. Sarigul-Klijn, A novel approach to extract power from free-flowing water and high-altitude jet streams, in Proceedings of the ASME energy sustainability conference, San Francisco, CA, ASME Energy Division Best Paper Award, ES2009-90146, 19–23 July 2009
2. R.E. Salomon, Process of converting wind energy to elemental hydrogen and apparatus therefor, US Patent 4,335,093, 15 June 1982
3. M. Meller, Wind-powered linear motion hydrogen production system, US Patent 7,146,918, 12 Dec 2006
4. A.R. Gizara, Turbine-integrated hydrofoil, US Patent 7,298,056, 20 Nov 2007
5. K.L. Holder, Regenerative generation method for hydrogen and oxygen, ammonia and methanol from wind energy into electrical energy, involving transforming energy and subsequent electrolysis of water with obtained electrical energy, German Patent DE 102007019027A1 (2007)
6. J. Kim, C. Park, Wind power generation with a parawing on ships, a proposal. J. Energy **35**, 1425–1432 (2010)
7. M.F. Platzer, N. Sarigul-Klijn, *Aerohydronautical Power Engineering – Is It the Key to Abundant Renewable Energy and Potable Water?* (University Readers, San Diego, 2012)
8. M.F. Platzer, N. Sarigul-Klijn, J. Young, M.A. Ashraf, J.C.S. Lai, Renewable hydrogen production using sailing ships. ASME J. Energy Resour. Technol. **136**(021203), 1–5 (2014)
9. M.F. Platzer, M. Lennie, D. M. Vogt, Analysis of the conversion of ocean wind power into hydrogen, in Proceedings of the world renewable energy, Congress, Perth, Australia, July 2013
10. M.F. Platzer, W. Sanz, H. Jericha, Renewable power via energy ship and Graz cycle, in Proceedings of the 15th international symposium on transport phenomena and dynamics of rotating machinery, ISROMAC-15, Honolulu, Hawaii, 24–28 Feb 2014
11. P.F. Pelz, M. Holl, M.F. Platzer, Analytical method towards an optimal energetic and economical wind-energy converter. J. Energy **94**, 344–351 (2016)
12. M. Holl, T. Janke, P.F. Pelz, M.F. Platzer, Sensitivity analysis of a techno-economic optimal wind-energy converter, in 2nd international conference on next generation wind energy, Lund, Sweden, 24–26 Aug 2016
13. M. Holl, M.F. Platzer, P.F. Pelz, Techno-economic comparison of renewable energy systems using multi-pole systems analysis, in 2nd International conference on energy production and management in the 21st century, Ancona, Italy, 6–8 Dec 2016
14. M. Holl, L. Rausch, P.F. Pelz, New methods for new systems – how to find the techno-economically optimal hydrogen conversion system. Int. J. Hydrogen Energy **42**, 22641–22265 (2017)

# Chapter 9
# Major Elements and Developmental Status of the Energy Ship Concept

The use of hydrokinetic turbines attached to cruising sailboats for the purpose of providing electric energy for navigation lights, GPS, radio, cell phones, and other devices is already well established. Dragged through the water behind the boat, a turbine manufactured by the Watt & Sea Company will produce about 130 watt at a boat speed of 5 knots, 300 watt at 7.5 knots, and 500 watt at 9 knots. Other companies selling turbines with similar power outputs are Swi-Tec, Sail-Gen, and H24O. The Watt & Sea Cruising 600 turbine is offered for 4158 Euro (exclusive VAT). It has a diameter of 240 mm.

A second important energy ship element is the conversion of the available onboard electricity into hydrogen. Fortunately, a complete hardware demonstration example for such a conversion is already available. Between 2004 and 2008, the Norsk Hydro Company developed and tested a full-scale wind-hydrogen energy system on the Norwegian island of Utsira [1]. The objective was to demonstrate how renewable energy can provide a safe, continuous, and efficient energy supply to remote areas. In this pilot project, 19 households with 235 inhabitants were supplied exclusively with wind turbine-generated energy. The excess wind power was used to produce 10 $Nm^3/h$ of hydrogen by means of a 48 kW electrolyzer. The hydrogen was compressed to 200 bar and stored in a tank. During times of insufficient or prohibitively strong wind activity, a 55 kW hydrogen internal combustion engine and a 10 kW proton-exchange membrane (PEM) fuel cell were used to convert the stored hydrogen into electric power. This project fully confirmed the technical feasibility of wind-to-hydrogen conversion, and much experience was gained during its 6 months stand-alone operation and maintenance of such a system. However, it was concluded that a permanent installation was not yet economically competitive with cable-supplied power from the mainland because the COE (cost-of-electricity) was 1.05 Euro/kWh.

The Utsira project inspired Chade et al. [2] to study the options available to provide Grimsey Island located 40 km north of Iceland with electric power for a community of 76 people. The peak electrical load is 175 kW with an average daily

M. F. Platzer, N. Sarigul-Klijn, *The Green Energy Ship Concept*, SpringerBriefs in Applied Sciences and Technology, https://doi.org/10.1007/978-3-030-58244-9_9

energy consumption of 2400 kWh. The monthly average wind speed varies between 7 and 9 m/s during September to April but decreases to 5 m/s during the summer months. They studied three options to replace the currently operating pure diesel system with either a combined wind-diesel system, a wind-hydrogen-diesel system, or a wind-hydrogen system. As on Utsira wind turbines were assumed to provide the electric power either for direct use or for producing hydrogen via electrolysis and tank storage for use during the low-wind summer months. They found that the wind-hydrogen-diesel system had the lowest operational costs of the three systems and all three systems reduced the system running costs during the postulated 20-year lifetime, but all required a substantial amount of initial investment due to the need for wind turbines, electrolyzer, fuel cell, hydrogen tank, and converter. A key aspect in sizing the system is the need to provide electrical power during the summer months which requires adequate hydrogen storage capacity or diesel generator capacity during periods of low wind speeds.

In 2015 Ostolski [3] investigated the possibility of replacing the wind turbines by energy ships to supply the Grimsey Island energy needs. He showed that it is reasonable to assume the availability of a 10 m/s average wind speed near the island except during the summer months, requiring again the need for hydrogen production and storage. Using the multi-pole systems analysis described in Chap. 23, he determined the need for a desalination power requirement of 0.6 kW, electrolysis power requirement of 929.8 kW, compressor power requirement of 19.6 kW, and a tank volume of 1625 kg. The energy ship needed to have a sail area of 1340 m$^2$ or of 660 m$^2$ if two ships are used. The hydrogen production rate had to be 6 kg/h or 3 kg/h, respectively. The hydrogen tank cost was assumed to be 325,000 Euro, and its O&M costs 3250 Euro/year; the fuel cell power was 175 kW at an investment cost of 87,850 Euro and O&M costs of 1820 Euro/year; the converter power was 175 kW at an investment cost of 120,000 Euro. The total investment costs became 4,558,791 Euro, comprising 50% for the ship, 7.7% for the turbine, 3% for the hydrogen compression, 0.5% for the desalination, 32.6% for the electrolyzer, and 6.2% for the hydrogen tank. Clearly, ship and electrolyzer are the major cost items for the implementation of the energy ship concept. The comparison with the three energy systems studied by Chade et al. [2] yields 0.409 Euro/kWh for the wind-diesel system, 0.278 for the wind-diesel-hydrogen system, 0.409 for the wind-hydrogen system, and 0.59 for the energy ship system. The hydrogen production cost became 14.38 Euro/kg hydrogen.

In this analysis the investment cost estimates for the ship, turbine, desalinator, electrolyzer, compressor, and hydrogen tank were based on data obtained from various manufacturers. Janke [4] provides the data sheets, and Ostolski [3] gives the following correlations, where I = Investment (Euro).

Ship           $I = 110.3 \, L^{2.63}$, L = ship length (m)
Turbine        $I = 46,285 \, A_T^{0.66}$, $A_T$ = turbine area (m$^2$)
Desalinator    $I = 17,211 \, P_D^{0.55}$, $P_D$ = desalinator power (kW)
Electrolyzer   $I = 19,299 \, P_E^{0.64}$, $P_E$ = electrolyzer power (kW)
Compressor     $I = 7259 \, P_C^{0.894}$, $P_C$ = compressor power (kW)
Tank           $I = p \, m_{H2}$, p = tank pressure (bar), $m_{H2}$ = hydrogen mass (kg)

Janke [4] gives the following data for sailboats, turbines, desalinators, electrolyzers, and compressors.

For example, the sailboat Cruiser 33, built by Bavaria, has a length of 9.99 m and a sail area of 51 m$^2$. The sales price is 66,900 Euro. According to the above correlation, the sales price is 46,928 Euro. The sailboat Easy 9.7, built by Bavaria, has the same length and sail area. It sells for 49,950 Euro. The difference in price is due to the cheaper materials used for Easy 9.7. The above correlations for the ship costs reflect boats with cheaper manufacturing costs.

For a turbine with a power output of 5 kW and a turbine area of 0.785 m$^2$, the sales price is 23,000 Euro, whereas for the same output and a turbine area of 1.767 m$^2$, the investment is 60,000 Euro. For a power output of 70 kW and a turbine area of 7.07 m$^2$, the investment is 140,000 Euro. Using the above correlation, this corresponds to 39,450, 67,393, and 168,290 Euro, respectively.

Janke [4] provides a list of 20 desalinators. The above correlation yields 216,669 Euro for a power input of 100 kW, 98,857 Euro for 24 kW, and 159,051 Euro for 57 kW. The actual list prices are 198,269, 79,112, and 164,634 Euro, respectively. Hence they compare reasonably well. The average price per kW is 2218 Euro.

Alkaline electrolyzer investment cost examples are 59219 Euro for 6 kW, 89,830 Euro for 12 kW, 134,837 Euro for 22 kW, and 210,526 Euro for 40 kW electrolyzers. The costs predicted from the correlation are 54,060, 94,670, 139,536, and 204,575 Euro, respectively. It should be noted that the price per kW therefore is 9870 Euro based on the price for the 6 kW electrolyzer. Janke lists prices per kW for other electrolyzers from 9870 Euro down to 1563 Euro per kW for a 960 kW electrolyzer. The Fraunhofer Institute [5] quotes an alkaline electrolyzer cost of approximately 1000 Euro/kW. This agrees with Schmidt et al. [46] who quote alkaline electrolyzer costs around 1000 Euro/kW, which are expected to drop to 750 Euro by 2030. The most recent review of the investment costs of electrolysis was given by Saba et al. [7] who project values of 397 Euro/kW by 2030.

It therefore appears that the multi-pole systems analysis carried out by Ostolski for Grimsey Island [3] which led to a hydrogen price of 14.38 Euo/kg hydrogen is very conservative. The electrolyzer costs represent a major part of the total investment costs. Hence considerable cost reductions can be expected as the electrolyzer costs are likely to decrease substantially in the near future.

PEM electrolyzer investment costs: Schmidt et al. [6] quote around 2000 Euro/kW which is expected to drop to 850 to 1650 Euro/kW by 2030. Saba et al. [7] project values of 955 Euro/kW by 2030.

Compressor investment cost examples are 60571 Euro for 11 kW and 128,571 Euro for 22.3 kW. The costs predicted by the correlation are 61,927 and 116,809 Euro, respectively. Ulleberg et al. [1] quote 5000 Euro/kW in their Utsira project. This cost compares well with the predicted cost of 5630 Euro/kW and 5238 Euro/kW.

Since the hydrokinetic turbine drag cannot be reduced below the minimum required to generate power, there is a great incentive to minimize the ship drag by the use of modern hydrofoil technology in combination with modern sail technology. Furthermore, personnel costs need to be minimized by operating the energy ships autonomously. The next step in the development of an economically

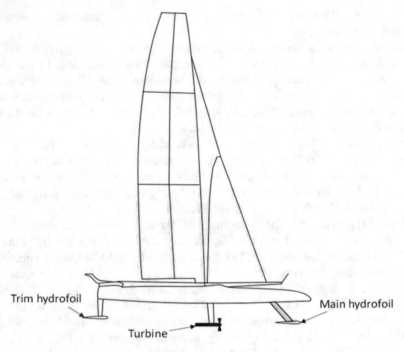

**Fig. 9.1** Unmanned autonomous hydrofoil energy ship [9]

competitive energy ship will be the design and testing of an autonomous hydrofoil energy ship, as shown in Fig. 9.1.

The energy ship concept, Fig. 8.2 permits two modes of operation. The electrical energy produced on board can either be stored in batteries or in compressed or liquefied hydrogen tanks. Battery storage avoids the need for desalination, electrolysis, and compression or liquefaction and therefore may be the most cost-effective method of delivering electric power for battery electric vehicles. For example, it is conceivable that autonomous sailboats could be used in wind-rich coastal areas or near islands to recharge car batteries during the night and to deliver these batteries in the morning as long as the travel distances can be kept small. Energy ships operating far from the coast, on the other hand, will have to store the energy in hydrogen tanks which are periodically delivered to the coast. This mode of operation involving relatively large travel distances was analyzed in some detail by Babarit et al. [8].

# References

1. O. Ulleberg, T. Nakken, A. Ete, The wind/hydrogen demonstration system at Utsira in Norway: Evaluation of system performance using operational data and updated hydrogen energy system modeling tools. Int. J. Hydrog. Energy, 1–12 (2009)

2. D. Chade, T. Miklis, D. Dvorak, Feasibility study of wind-to-hydrogen system for Arctic remote islands – Grimsey island case study. Renew. Energy **76**, 204–211 (2015[21)

3. A. Ostolski, *Energy supply optimization for potential estimation of an innovative wind-energy converter,* Research Report S 216, Fluid Systems Technology, Technical University Darmstadt, 23 November 2015 (in German)

4. T. Janke, *Sensitivity analysis of a techno-economic optimal wind-energy converter,* Research Report S 240, Fluid Systems Technology, Technical University Darmstadt, 19 February 2016 (in German)

5. T. Smolinka, M. Guenther, J. Garche, *Stand und Entwicklungspotential der Wasserelektrolyse zur Herstellung von Wasserstoff aus regenerativen Energien,* Fraunhofer NOW-Studie, 22.12.2010

6. O. Schmidt et al., Future cost and performance of water electrolysis; an expert elicitation study. Int. J. Hydrog. Energy **42**, 30470–33049 (2017)

7. S.M. Saba, M. Mueller, M. Robinius, D. Stolten, The investment costs of electrolysis – A comparison of cost studies from the past 30 years. Int. J. Hydrog. Energy **43**, 1209–1223 (2018)

8. A. Babarit, J.C. Gilateaux, G. Clodic, M. Duchet, A. Simoneau, M.F. Platzer, Techno-economic feasibility of fleets of far offshore hydrogen-producing wind energy converters. Int. J. Hydrog. Energy **43**, 7266–7289 (2018)

9. M.F. Platzer, N. Sarigul-Klijn, Storable energy production from wind over water. J. Energy Power Technol. **2**(2) (2020). https://doi.org/10.21926/jept2002005

# Chapter 10
# Comparison of the Wind-over-Water with the Wind-Water-Solar Concept

When comparing and ranking the energy ship (WoW) concept with the wind-water-solar (WWS) concept using the previously formulated ranking criteria, the following observations can be made:

(a) Renewable energy availability: the amount of energy contained in the winds over the oceans far exceeds that on land or close to shore.

(b) Power density: the ocean wind power density exceeds that of the winds over land and close to land.

(c) Land use: the energy ship (WoW) system requires no land use for the energy production.

(d) Capacity factor: the capacity factor of energy ships can be expected to be twice that of land-based wind turbines.

(e) Economic viability: Pelz et al. [1], Holl et al. [2], and Ostolski [3] performed an economic analysis of the energy ship (WoW) system using available cost data about conventional displacement ships, hydrokinetic turbines, desalinators, electrolyzers, compressors, and hydrogen tanks. They arrived at a price of approximately $ 15 per kg of hydrogen. This price is three to four times higher than the current market price for hydrogen produced by steam reforming. It is well known that the cost of wind turbines dropped by about 70% since the introduction of wind turbines in the 1970s. Hence it is reasonable to expect that a similar decline can be expected as soon as the various components of the energy ship (WoW) system start to be produced at a larger scale and as the components start to be optimized. For example, the replacement of displacement by hydrofoil-borne ships is likely to double or triple the power production rate due to the speed increases made possible by the greatly reduced hydrodynamic ship drag. Hence it is likely that energy ships will be able to compete economically with offshore wind turbines within a few years.

(f) Energy storability: energy ships are forced to store the produced energy in the form of compressed or liquefied hydrogen or in electric batteries.

© The Editor(s) (if applicable) and The Author(s), under exclusive license to
Springer Nature Switzerland AG 2021
M. F. Platzer, N. Sarigul-Klijn, *The Green Energy Ship Concept*, SpringerBriefs in
Applied Sciences and Technology, https://doi.org/10.1007/978-3-030-58244-9_10

(g) Socio-political acceptability: it is our view that the energy ship concept offers a distinct advantage over the WWS system. The placement and operation of land- and offshore-based power production systems require a lengthy approval process by various local, state, and federal agencies, thus providing many opportunities to various interest groups to delay or scuttle renewable energy projects. In contrast, energy ships operating in international waters are subject only to the regulations covering ship traffic on the high seas. As a consequence, it is reasonable to expect that the energy ship technology will experience a very rapid development as soon as various companies, organizations, and governments recognize its potential for a rapid transition to emission-free power production at an acceptable cost.

# References

1. P.F. Pelz, M. Holl, M.F. Platzer, Analytical method towards an optimal energetic and economical wind-energy converter. J. Energy **94**, 344–351 (2016)
2. M. Holl, L. Rausch, P.F. Pelz, New methods for new systems – How to find the techno-economically optimal hydrogen conversion system. Int. J. Hydrog. Energy **42**, 22641–22654 (2017)
3. A. Ostolski, *Energy Supply Optimization for Potential Estimation of an Innovative Wind-Energy Converter*. Research Report S 216, Fluid Systems Technology, Technical University Darmstadt, 23 November 2015 (in German)

# Chapter 11
# Sustainable Aviation

In 2016 we drew attention to the possibility of using energy ships to extract both hydrogen and carbon dioxide from seawater for the purpose of producing sustainable aviation fuels [1]. In recent years, the US Naval Research Laboratory studied the possibility of replacing the conventional jet fuel by a synfuel manufactured from hydrogen and carbon dioxide extracted from seawater. Willauer et al. [2] performed a cost/benefit and energy balance analysis of this type of synthetic jet fuel production using electric power delivered from either nuclear or from ocean thermal energy conversion (OTEC) power plants. In order to produce 100,000 gallons per day of jet fuel, 443,900 m$^3$/day $CO_2$ and 1,373,600 m$^3$/day of hydrogen are required. The production of this amount of hydrogen requires the processing of 1082 m$^3$/day of water and the delivery of 246 MW of electric power. They concluded that this type of jet fuel could be produced from seawater at a cost of $ 3 to $ 8/gal.

Very recently, Willauer et al. [3] succeeded in developing an electrolytic cation exchange module which enables the extraction of large quantities of carbon dioxide from seawater and to simultaneously produce hydrogen in quantities and ratios needed for the jet fuel synthesis. The key for the adoption and scale-up of this technology is the energy efficiency and the carbon dioxide production efficiency. Energy ship produced electricity therefore can be an important contributor to the economic production of synthetic jet fuel.

M. F. Platzer, N. Sarigul-Klijn, *The Green Energy Ship Concept*, SpringerBriefs in
Applied Sciences and Technology, https://doi.org/10.1007/978-3-030-58244-9_11

# References

1. M.F. Platzer, N. Sarigul-Klijn, Carbon-neutral jet fuel production from seawater. Int. J. Sustain. Aviat. **2** (2016)
2. H.D. Willauer, D.R. Hardy, K.R. Schultz, F.W. Williams, The feasibility and current estimated capital costs of producing jet fuel at sea using carbon dioxide and hydrogen. J. Renew. Sustain. Energy **4**, 033111 (2012). https://doi.org/10.1063/1.4719723
3. H.D. Willauer, F. DiMascio, D.R. Hardy, F.W. Williams, Development of an electrolytic cation exchange module for the simultaneous extraction of carbon dioxide and hydrogen gas from natural seawater. Energy Fuel **31**, 1723–1730 (2017)

# Chapter 12
# Proposal for a Global Renewable Energy Production and Storage Initiative

A review of the vast literature on the danger posed by climate change shows an amazing absence of the global engineering community in proposing specific approaches to implement substantial global carbon dioxide reductions by a specified deadline. The Saturn-Apollo project is often cited as an example to be adopted for the formulation of a Global Apollo Project for renewable energy production. Unfortunately, the Global Apollo Program proposed by King et al. [1] lacks the specificity to convince the politicians and the general public that solutions are possible.

We therefore suggest that there is an urgent need for the establishment of a Global Engineering Council for the purpose of soliciting proposals and conducting a comparative evaluation of these engineering proposals to achieve the goal of global renewable global power production and storage no later than by mid-century [2]. This Council should also develop criteria for the purpose of ranking the various proposals as to their technical, economic, and socio-political feasibility. Furthermore, it should summarize the methodology used for their rankings in generally understandable terms and disseminate the results of its deliberations to various governmental entities and to the general public through various media outlets. Our own views on the available engineering options are given in reference [3].

## References

1. D. King, J. Browne, R. Layard, G. O'Donnell, M. Rees, N. Stern, A. Turner, *A Global Apollo Programme to Combat Climate Change*. Center for Economic Performance, The London School of Economics and Political Science, 2015
2. M.F. Platzer, On the need for a global engineering initiative to mitigate climate change. Int. J. Energy Prod. Manag. **1**(2), 155–162 (2016)
3. M.F. Platzer, N. Sarigul-Klijn, *Engineering Options for an Emission-Free Global Economy by 2050*. Proceedings of the ASME 2016 International Mechanical Engineering Congress and Exposition IMECE2016–66345, November 11–17, 2016, Phoenix, Arizona

M. F. Platzer, N. Sarigul-Klijn, *The Green Energy Ship Concept*, SpringerBriefs in Applied Sciences and Technology, https://doi.org/10.1007/978-3-030-58244-9_12

# Chapter 13
# Summary and Outlook

In this book we have attempted to show that the increasingly urgent challenge of eliminating the carbon dioxide emissions by mid-century can be met by dropping two assumptions which have been regarded as self-evident by most analysts. We argue that any comprehensive analysis has to start with a ranking of the global renewable energy resources available on the planet. Given the fact that 70% of the planet's surface is covered by water and that most of the global wind energy can be found over water, the question needs to be answered whether the "wind-over-water" energy resources can be tapped for conversion into storable energy. The second assumption which needs to be re-examined is the widely held view that any power plant by its very nature has to be large and sturdy and therefore requires a firm foundation which, in turn, makes it immobile. Our analysis shows that both assumptions do not withstand close scrutiny. The sailing ship technology which has been successfully used for many centuries to transport people and commercial goods over large distances deserves consideration for energy conversion purposes for four major reasons:

(a) It allows the extraction of energy from an essentially inexhaustible energy reservoir.
(b) It enables the development of an entirely new energy production method.
(c) It moves the energy production into areas which are of little use for other human activities.
(d) It solves the energy storage problem.

It is tempting to believe that these "rational" arguments should be sufficient to initiate a rapid global conversion process to renewable energy production. Unfortunately, human history teaches a different lesson. Massive changes occur only in response to circumstances which force the acceptance of new economic and/or socio-political conditions. The fear of overthrow of an established socio-political order is a very effective motivator for groups of people to rally together. Real or perceived military threats by one country (or group of countries) against another

M. F. Platzer, N. Sarigul-Klijn, *The Green Energy Ship Concept*, SpringerBriefs in
Applied Sciences and Technology, https://doi.org/10.1007/978-3-030-58244-9_13

country (or group of countries) bring about rapid changes in both the victorious and the defeated countries. Unfortunately, climate change is still largely perceived as a threat which does not require an emergency response. As a consequence, the inherent system inertia and the existing commercial and political special interests largely succeed in most countries to prevent or delay effective action.

However, there is one "tipping point" which may soon be reached. Young people may soon become convinced that the acceptance of "business as usual" is endangering their future to a degree which will motivate them to demand a global "Energy Apollo Program." We are confident that most secondary and university students can be taught that "nature cannot be fooled." They will start understanding that the Laws of Nature cannot be violated and ignored indefinitely. They will start understanding that they are passengers in a single spaceship which needs to be repaired in order to remain a livable home and that it is therefore in their interest to rally together and to demand action.

For this reason, it will be crucial to teach young people how to distinguish scientific information from ideological misconceptions and outright demagoguery. They need to learn how to look for the empirical information which led to the formulation of the Laws of Nature, and they need to understand that the empirical database will never be complete and totally free of uncertainty. While certain scientific disciplines have matured to the point of universal acceptance (examples are Newton's or Maxwell's equations), other areas are still so complex that they cannot be reduced to the solution of a few equations. The prediction of the effect of greenhouse gas emissions on the climate is one such area. Climate change skeptics will always be able to point to gaps and inconsistencies in the available data which forces the climatologists to formulate the predictions of future trends in terms of probabilities rather than certainties.

Therefore, young people need to be taught about decision-making in the face of uncertainty. It will help them to counter the arguments of the climate skeptics and demagogues by pointing to decision-making in aeronautical engineering. Aeronautical engineers are still unable to predict the failure of an aircraft due to the formation of fatigue cracks in the airplane wings, fuselage, and engine blades. They have learned that it is necessary to pay attention to the formation of small cracks. They know that the airplane is able to fly for an extended period of time with such cracks, but they also know that the probability of failure increases with time. Therefore, they have learned to insist on rigorous airplane inspection, maintenance, and repair rules. Airplane passengers need to be sure that their plane has been properly maintained. Spaceship Earth passengers are entitled to the maintenance and repair of their ship in the face of clear evidence that the air-conditioning system of spaceship Earth is starting to fail!

Repair of spaceship Earth therefore requires the cessation of carbon dioxide emission into the atmosphere. The young people need to be taught that it is technically feasible to achieve this goal by mid-century. It is the responsibility of the global engineering community to generate enthusiasm for the attainment of this goal. This will require the teaching of climate science and renewable energy engineering principles at the secondary and tertiary levels. Most importantly, it will require active participation of young people, in particular of science and engineering students, in projects which make a positive contribution to climate change alleviation.

It is our view that young people always respond quite enthusiastically to well-posed technical challenges. Indeed, young people, such as the Wright brothers [1], Wernher von Braun [2], Sergei Korolev [3], Hans von Ohain [4], and many others, were fascinated by the challenge of achieving powered flight or developing turbojet or rocket engines. They started their projects at a very small scale with little or no support. The Wright brothers built their airplane without any outside support and barely managed to fly the length of a football field. Wernher von Braun joined a group of rocket enthusiasts during his student years which had virtually no support and achieved rocket flights of only a few miles, yet 30 years later, he started the construction of the rocket which enabled the landing of two astronauts on the moon in 1969. At about the same time, Sergei Korolev joined a group of rocket enthusiasts in Moscow to experiment with small rockets. He succeeded in building the rocket which put the first satellite into earth orbit, thereby starting the "space age." Hans von Ohain was a bit luckier. Shortly after obtaining his PhD, he was able to persuade Ernst Heinkel, the owner of the Heinkel Aircraft Company, to give him a few technicians for the demonstration of a small jet engine. We mention these histories to make it clear to today's young people that new technological developments often start as very small ventures with little or no support, yet occasionally grow amazingly fast. It is often forgotten that 15 years after the flight of the Wright airplane, at the end of World War I, well over 200,000 airplanes had been built. Wernher von Braun and his team needed only 10 years to demonstrate the first supersonic missile, and the first operational jet fighter airplane, the Messerschmitt 262 airplane, flew only 5 years after von Ohain's first demonstration of his jet engine in the Heinkel 178 airplane in 1939. It took only 13 years for the first commercial jet airplane, the Comet [5], to enter service.

We mention these facts to impress upon young people that they do not need to wait for the government or other large organizations "to do something" about climate change. Small-scale demonstration projects can generate enough interest to attract funding for larger-scale developments. For this reason, we pose the following challenge.

Think about building a vehicle which can convert the wind energy contained in winds over water into electric energy. Consider the fact that a water turbine can produce power at the rate of $(500) V^3 A$ $(0.593)$ where $V$ is the water speed, $A$ is the turbine disk area, and the factor $(0.593)$ represents the Betz factor (see Chap. 20). This formula shows that a significant amount of power can be generated if the water speed is high because a doubling of the water speed increases the power output by a factor of eight. Hence a small water turbine can generate a significant amount of power. Therefore, think about moving a small turbine through stationary ocean or lake water at as high a speed as possible. To this end you need to capture a certain amount of wind to convert the wind power into thrust by means of sails, wings, or other mechanisms so that you can push the turbine through the water. You may call this vehicle a sailboat or, more generally, an "air-sea interface vehicle," since a conventional sail may not necessarily provide the optimum wind to thrust conversion mechanism. The challenge is to find the vehicle configuration which maximizes the water turbine power output. Note that it is entirely sufficient to demonstrate a small-scale vehicle. As shown by the above mentioned aviation pioneers, the building of larger vehicles follows "automatically."

In summary, it is the objective of this book to convince the readers that climate change demands the conversion of the fossil-based global economy to an environmentally sustainable economy. To this end, we show that current renewable energy production methods are unlikely to reach the goal of a virtually complete transition by mid-century. We alert the readers to the fact that there are untapped energy resources on our planet in the form of the ocean winds which can be converted into storable energy by means of autonomous specially designed air-sea interface vehicles. Also, we draw attention to the fact that these vehicles present a design challenge similar to the challenge of designing and demonstrating air and space vehicles which made possible the "conquest of space and time" by enabling intercontinental flight and instantaneous worldwide communication. Therefore, our book is primarily aimed at engineering professionals and students to stimulate their interest in a new engineering discipline which combines aeronautical, hydronautical, and power engineering and which we therefore called "aerohydronautical power engineering" in 2012 [6]. Its purpose is to conquer the "energy barrier" by constructing and operating fleets of energy ships. We contend that autonomous hydrofoil-borne ships may well become economically competitive in the near future. Therefore, the construction of some 500,000 ships per year for 30 years can make possible the transition to an emission-free global economy by mid-century. We hope that this prospect will also convince a wider readership that an "Energy Apollo Program" involving aerohydronautical power engineering is technically and economically feasible [7].

# References

1. P. Scott, *The Pioneers of Flight* (Princeton University Press, 1999)
2. M.J. Neufeld, *Von Braun* (Alfred A Knopf, New York, 2007)
3. J. Harford, *Korolev* (Wiley, 1997)
4. M. Conner, *Hans von Ohain* (American Institute of Aeronautics & Astronautics, 2001)
5. S.H. Verhovek, *Jet Age, the Comet, the 707, and the Race to Shrink the World* (Penguin Group, New York, 2010)
6. M.F. Platzer, N. Sarigul-Klijn, *Aerohydronautical Power Engineering – Is It the Key to Abundant Renewable Energy and Potable Water?* (University Readers, San Diego, 2012)
7. M.F. Platzer, N. Sarigul-Klijn, Transition to an emission-free global economy by mid-century using energy ships. Int. J. Clim. Change: Impacts Responses **10**(4), 41–52 (2018)

# Part II
# Technical Aspects

# Chapter 14
# Energy and Power Fundamentals

*Energy*: the ability to do work, also a quantity or volume of fuel. The rate at which
  energy is used is called power. Energy is measured in joules. For example, an
  energy supply providing one joule per second gives one watt of power. Thus a
  60-watt bulb provides 60 joules each second.

1 kilowatt-hour (kWh) = 3,600,000 joules energy provided over 1 h
1 kilowatt-hour (kWh) = 3600 kilo joules = 3.6 mega joules

*Energy density*. the amount of energy that can be contained in a given unit of
  volume, area, or mass. It is typically expressed in joules per kilogram.
*Joule (J)*: a unit of energy, whereby 1 joule = 1 watt-second; the energy exerted by
  the force of 1 Newton acting to move a mass of 1 kg a distance of 1 m.
*Power*: the rate at which energy is used or at which work gets done. Power is
  measured in watts, after the inventor of the modern steam engine, the Scotsman
  James Watt (1736–1819).
*Power density*: the amount of power that can be harnessed in a given unit of volume,
  area, or mass. The comparison of renewable energy sources to other sources is
  typically done in watts per square meter.
*Watt*: a unit of power. By definition, 1 watt = 1 joule per second.
*Energy conservation law*: a fundamental law of physics.
*Capacity factor*: actual amount of energy produced per year divided by the maxi-
  mum possible energy.

© The Editor(s) (if applicable) and The Author(s), under exclusive license to
Springer Nature Switzerland AG 2021
M. F. Platzer, N. Sarigul-Klijn, *The Green Energy Ship Concept*, SpringerBriefs in
Applied Sciences and Technology, https://doi.org/10.1007/978-3-030-58244-9_14

# Chapter 15
# Hydrogen Characteristics

Hydrogen is designated by the symbol H. It has the atomic number 1 and is the lightest element in the periodic table. It is the most abundant element in the universe and was first identified and isolated by H. Cavendish in 1766. At room temperature it is colorless, odorless, and not very reactive, unless activated by an appropriate catalyzer. At high temperature it is highly reactive. Although it is diatomic, molecular hydrogen dissociates into free atoms at high temperatures. The density of atomic hydrogen is 0.08988 $kg/m^3$. Compressed hydrogen gas has a density of 23–27 g/l at a pressure of 350 bar and 38–40 g/l at a pressure of 700 bar.

Pure hydrogen does not exist in appreciable amounts. Instead it is always connected to other elements, like carbon in plants, petroleum, and natural gas or in water where it is attached to oxygen. Therefore, an energy source is needed to separate hydrogen from its partner. This can be done in several ways, for example, from natural gas, water, biomass, or coal, and a number of energy sources can be used to accomplish the separation, such as wind, solar, coal, natural gas, and nuclear. Once extracted, hydrogen can provide heat and electricity through combustion or reaction in a fuel cell. The oxidation of hydrogen follows the reaction: $2H_2 + O_2 \rightarrow 2H_2O$. The enthalpy of combustion for hydrogen is approximately 141 mega joules per kilogram when the product is liquid water, otherwise known as the higher heating value (HHV). The enthalpy of combustion drops to 121 mega joules per kilogram when the product is water vapor, otherwise known as the lower heating value (LHV). The enthalpy of combustion for hydrogen nearly triples that of natural gas, propane, gasoline, diesel fuel, and jet fuel.

Hydrogen is extremely flammable. The gas mixes well with air and explosive mixtures are easily formed. The gas is lighter than air. Heating may cause violent combustion or explosion. High concentrations in the air may cause a deficiency of oxygen, leading to unconsciousness or death. There are no odor warnings if toxic concentrations of hydrogen are present.

© The Editor(s) (if applicable) and The Author(s), under exclusive license to
Springer Nature Switzerland AG 2021
M. F. Platzer, N. Sarigul-Klijn, *The Green Energy Ship Concept*, SpringerBriefs in
Applied Sciences and Technology, https://doi.org/10.1007/978-3-030-58244-9_15

Unit conversion for hydrogen: $Nm^3$ (normal cubic meter) to kg and kg to $Nm^3$

$1\ Nm^3 = 0.08988$ kg
$1$ kg $= 11.126\ Nm^3$

Hydrogen properties at 1 atm normal conditions (1 atm and 0 °C)

Hydrogen   2.016 g/mol      0.08988 $kg/m^3$   LHV (lower heating value)
                                                                        241.8 kJ/mol

Oxygen       31.999 g/mol    1.429 $kg/m^3$

Hydrogen has the highest gravimetric energy density compared to other energy sources, but its volumetric energy density is very low. Therefore, the only way to compensate for the low volumetric energy density of hydrogen is to either compress the gas, liquefy it, or bond hydrogen into another substance. Compression is a straightforward method for increasing the volumetric energy density for short periods of time for two key reasons. First, hydrogen is a gas under practical temperatures and pressures. Its critical temperature, −239.96 °C, and pressure, 12.98 atm, necessitate the use of cryogenic refrigeration to bring hydrogen into liquid form. Second, hydrogen is most commonly used as a fuel under atmospheric temperatures and pressures. Storage in the same form in which the hydrogen will ultimately be used will not require additional active subsystems to maintain the storage temperature and pressure.

# Chapter 16
# Hydrogen Production Methods

## 16.1 Steam Reforming

Most of the hydrogen used in industry is produced by steam-methane reforming where high-temperature steam is used to produce hydrogen from a methane source, such as natural gas. Methane reacts with steam under 3–25 bar pressure in the presence of a catalyst to produce mostly hydrogen and carbon monoxide. In a subsequent "water-gas shift reaction," carbon monoxide and steam are reacted using a catalyst to produce carbon dioxide and additional hydrogen. Carbon dioxide then is removed by "pressure-swing adsorption" to produce pure hydrogen.

## 16.2 Renewable-Based Methods: Electrolysis

The major disadvantage of hydrogen production by steam reforming of natural gas or liquid hydrocarbons is the emission of greenhouse gases. Therefore, the only major emission-free hydrogen generation method is by electrolysis to split water into hydrogen and oxygen by passing an electric current through the water. In 1800 William Nicholson and Anthony Carlisle first demonstrated this method which was perfected by Zenobe Gramme in 1869.

Electrolysis is the process of disassociating water molecules into hydrogen and oxygen gas by passing a direct electric current through two electrodes in water. The basic components of an electrolytic cell consist of an anode, cathode, diaphragm membrane, electrolyte, and a direct current (DC) power supply. Figure 16.1 shows the basic scheme of an alkaline water electrolysis system. At the cathode, hydrogen ions ($H^+$) consume the electrons provided by the DC power supply and form hydrogen gas. To satisfy Kirchhoff's current law, the leftover hydroxide ions must transfer through the electrolyte to the anode and give away the electrons to return to

M. F. Platzer, N. Sarigul-Klijn, *The Green Energy Ship Concept*, SpringerBriefs in
Applied Sciences and Technology, https://doi.org/10.1007/978-3-030-58244-9_16

**Fig. 16.1** Basic scheme of
an alkaline water
electrolyzer

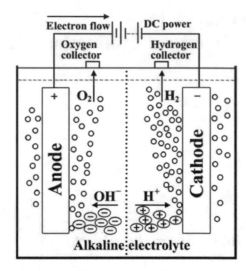

the positive terminal of the power supply to complete the circuit. There is a diaphragm membrane in between the two electrodes. The purpose of the membrane is to prevent the electrodes from touching each other to create a short. Its other function is to stop the recombination of oxygen and hydrogen and selectively allow hydroxide ions to pass through. By giving away the electrons, the hydroxide ions oxidize to oxygen gas and water.

Currently, three types of low-temperature industrial electrolyzers are produced – unipolar electrolyzers, bipolar electrolyzers, and solid polymer electrolyte electrolyzers. Alkaline electrolyzers use an aqueous solution of potassium hydroxide (KOH) because of KOH's high conductivity and because the oxygen evolution reaction has the least energy loss in this solution. These electrolyzers do not require precious metals but use only nickel electrodes. Unipolar electrolyzers have electrodes connected in parallel with a single bus bar, connecting all the anodes and another connecting all the cathodes. The bipolar alkaline electrolyzer cells are connected in series so that hydrogen is produced on one side of each cell and oxygen on the other. The major difference between alkaline water and solid polymer electrolyzers is the electrolyte. The solid polymer electrolyzer, also referred to as proton exchange membrane or polymer electrolyte membrane electrolyzer, uses a solid polymer called the polymer electrolyte membrane (PEM) as the electrolyte. Because the solid polymer electrolyzer requires the use of an electrode that is embedded with an exotic metal such as platinum and iridium as a catalyst, the cost of a solid polymer electrolyzer is on average eight times more expensive than an alkaline water electrolyzer. Alkaline electrolyzers split water at 0.1–0.4 A/cm$^2$ in a KOH electrolyte while using a microporous divider known as a diaphragm to separate the two electrodes. PEM electrolyzers are based on a "zero-gap" membrane electrode assembly design in which proton-conducting solid polymer electrolytes such as Nafion are sandwiched between porous electrode layers. This design enables higher operating current densities (0.6–2 A/cm$^2$) than in alkaline electrolyzers. In both types of electrolyzers, the membrane and diaphragm permit the transport of ions between the electrodes while simultaneously separating the hydrogen and oxygen to prevent the formation of an explosive mixture. The need for membrane functionality and durability in PEM electrolyzers calls for the use of corrosion-resistant titanium. Diaphragms in alkaline electrolyzers are simpler and less expensive, but they are still susceptible to blockage by impurities.

There is therefore a great incentive to develop electrolyzers without membranes or diaphragms where forced fluid flow (advection) and/or buoyancy forces are used to separate the oxygen and hydrogen products before they can cross over to the opposing electrode. Such membraneless electrolyzers have the potential advantages of decreased capital costs and increased durability with long operating lifetimes, higher tolerance to impurities, greater resilience to extreme operating conditions, and ability to use a greater variety of electrolytes. As reviewed recently by D.V. Esposity [1], the membraneless electrolyzer technology has the potential to become a key element in the transition to emission-free energy production.

Direct seawater electrolysis is technically feasible. However, the biggest problem when using salt water in an electrolyzer with a high current density is the creation of chlorine at the anode. This can be alleviated by the use of special anode coatings, but the energy requirement is typically twice that of fresh water electrolyzers. Therefore, water desalination by reverse osmosis before electrolysis is the preferred option.

## 16.3  Representative Commercial Electrolyzers

The Norwegian company NEL Hydrogen offers the high-pressure alkaline NEL P-60 electrolyzer. It is a compact turnkey hydrogen plant that can be delivered in a 20 foot container or skid-mounted for indoor installation. The dimensions of the container are $6.1 \times 2.5$ m $\times 2.6$ m and of the electrolyzer skid 3.5 m $\times 2.3$ m $\times 2.2$ m. It produces hydrogen with an outlet pressure of 15 bar and a maximum capacity of 60 Nm$^3$ (5.5 kg) per hour from one single electrolyzer cell stack. The power consumption per Nm$^3$ of hydrogen is 4.9 kWh. The feed water flow rate is approximately 0.9 l per Nm$^3$ of hydrogen. The flow rate of the cooling water is approximately 15 m$^3$ per hour at maximum capacity. The electrolyte is 32% KOH aqueous solution. Its distinguishing feature from other electrolyzers is the flexibility of production down to 10% of installed capacity.

The French company SAGIM S.A. offers electrolyzers type BP-MP 1000/5000 that can deliver 1–5 Nm$^3$ (0.45 kg) of hydrogen per hour at a maximum output pressure of 10 bar requiring 5 kWh/Nm$^3$. The dimensions are 1950 (H) $\times$ 950 (w) $\times$ 2500 to 4000 mm (L). The weight varies between 1000 and 2500 kg, and the storage capacity is 6 to 24 Nm$^3$.

The Canadian company Hydrogenics in Mississauga, Ontario, offers the electrolyzer HySTAT 10. It produces 8.6 to 21.5 kg/day and requires 4.9 kWh per Nm$^3$. It requires 1.5–2.0 l of water into the reverse osmosis system per Nm$^3$ of hydrogen. For demineralized water it requires less than 1 l/Nm$^3$ of hydrogen. The electrolyzer requires a volume of $1.7 \times 1.85 \times 2.6$ m and the power/control cabinet $1 \times 0.5 \times 2.1$ m. The weights are 1400 kg and 1200 kg, respectively. Other electrolyzers are HySTAT 15, HySTAT 30, and HySTAT 60.

## 16.4  Electrolyzer Cost Information

The most recent comprehensive cost and performance studies of water electrolysis were published by Schmidt et al. [2] and Saba et al. [3] where the current alkaline electrolyzer costs are quoted as around 1000 euro/kW, which are expected to drop to 750 euro and 397 euro, respectively, by 2030. For PEM electrolyzers, Schmidt et al. [2] quote a current cost of around 2000 euro/kW which is expected to drop to 850 to 1650 euro/kW by 2030. Saba et al. [3] project a decrease to 955 euro/kW by 2030.

## References

1. D.V. Esposito, Membraneless electrolysers for low-cost hydrogen production in a renewable energy future. Joule **1**, 651–658 (2017)
2. O. Schmidt et al., Future cost and performance of water electrolysis; An expert elicitation study. Int. J. Hydrog. Energy **42**(52), 30470–30492 (2017)
3. S.M. Saba, M. Mueller, M. Robinius, D. Stolten, The investment costs of electrolysis – A comparison of cost studies from the past 30 years. Int. J. Hydrog. Energy **43**, 1209–1223 (2018)

# Chapter 17
# Seawater Desalination

Several methods are available to desalinate seawater: reverse osmosis, vacuum distillation, multistage flash distillation, multiple-effect distillation, vapor compression distillation, freeze thaw, solar evaporation, electrodialysis reversal, membrane distillation, and wave-powered distillation.

Reverse osmosis is the commonly used system for large-scale desalination. Information about seawater desalination power consumption is provided in reference [1]. Typically, for a 500 kW power input into an electrolyzer, a fresh water flow of 2.16 m³/day is required, whereas for 5 MW, 21.6 m³/day are required.

The company Pure Aqua, Inc. provides the following information on their reverse osmosis system. For a fresh water flow of 2.6 m³/day, the required volume is 0.21 m³, the weight is 104 kg, and the power consumption is 1.5 kW. For a fresh water flow of 20.8 m³/day, the volume is 1.7, the weight is 386 kg, and the power consumption is 7.5 kW.

The reverse osmosis cost is very small compared to the other energy conversion equipment cost. According to Janke [2], the average cost is 2218 Euro/kW, based on 20 desalinators with power inputs between 24 and 322 kW. Ostolski [3] used this data to derive the correlation between the cost I (euro) and the power P(kW) as $I = 17{,}211\ P\ \exp(0.55)$.

# References

1. Water Use Association Desalination Committee, Seawater Desalination Power Consumption, White Paper, November 2011
2. T. Janke, *Sensitivity Analysis of a Techno-economic Optimal Wind-Energy Converter*. Research Report S 240, Technical University Darmstadt, 19 February 2016
3. A. Ostolski, *Energy Supply Optimization for Potential Estimation of an Innovative Wind Energy Converter*. Research Report S 216, Technical University Darmstadt, 23 November 2015

© The Editor(s) (if applicable) and The Author(s), under exclusive license to          63
Springer Nature Switzerland AG 2021
M. F. Platzer, N. Sarigul-Klijn, *The Green Energy Ship Concept*, SpringerBriefs in
Applied Sciences and Technology, https://doi.org/10.1007/978-3-030-58244-9_17

# Chapter 18
# Energy Storage Systems

The selection of an energy storage system depends on the required discharge time and energy density. There are three categories of applications: power quality, bridging power, and energy management. Power quality applications are employed to provide transient stability and frequency regulation. These applications require rapid response, typically within a fraction of a second. Flywheels, capacitors, and superconducting magnetic energy storage (SMES) are suitable for power quality applications. Bridging power applications are employed as reserve to bridge the gap between generation and demand due to forecast uncertainty and unit commitment error. The response time for bridging power applications is relatively longer than power quality applications and ranges from seconds to minutes. Discharge times can be up to an hour. Several battery technologies, such as lead-acid, nickel-cadmium, nickel-metal hydride, and lithium-ion batteries, are suitable for bridging applications. Energy management applications are employed to store energy at a time of low demand and discharge the energy at a time of high demand. These applications generally require a continuous discharge time of several hours or more. High-energy density batteries, pumped hydro, compressed air, and hydrogen energy storage are suitable for energy management applications. Table 18.1 summarizes the three categories of energy storage applications.

**Table 18.1** Three categories of energy storage

| Category | Discharge time required | Technology |
| --- | --- | --- |
| Power quality | Seconds to minutes | Flywheels, capacitors, and superconducting magnetic energy storage |
| Bridging power | Minutes to ~1 hour | Batteries |
| Energy management | Hours | High-energy density batteries, pumped hydro, compressed air, and hydrogen and ammonia energy storage |

M. F. Platzer, N. Sarigul-Klijn, *The Green Energy Ship Concept*, SpringerBriefs in
Applied Sciences and Technology, https://doi.org/10.1007/978-3-030-58244-9_18

A very comprehensive review of various energy storage systems is given by Ausfelder et al. [1].

## 18.1   High-Energy Density Batteries Storage System

Prior to 2009 the technologies for high-density batteries were sodium based, nickel based, or lead-acid based. Since then, the advancement of lithium ion technology has increased the utilization of batteries in grid scale energy storage systems. The lithium ion battery is favored over other batteries because of its high-energy density. With a nominal voltage of 3.7 V, it is much higher than many other batteries. That means fewer cells are needed to produce the same amount of power. In addition, lithium ion batteries have an efficiency of 85–95% and an expected lifetime of 10–15 years or 2000–3000 cycles.

In 2012 Watson and Frank [2] proposed to use the batteries of electric vehicles for energy storage. To this end, they studied the energy needs on the Hawaiian Islands. In 2012 the total energy need was 46.7 million kWh/day, requiring an average power supply of 1.95 million kW and a peak power supply of 3.3 million kW. The total number of vehicles was 1.13 million vehicles which used about 8 million kWh/day, and homes, businesses, and industries used about 38.7 million kWh/day, so that the total use was 46.7 million kWh/day. Since, for example, the Chevrolet Volt car has a 15 kWh battery, but only 6 kWh are normally consumed for daily driving needs, at least 7 kWh in 1.13 million vehicles are available to provide energy storage of 8 million kWh/day. This energy therefore can be used to compensate for the fluctuation of renewable energy supplied by wind turbines and solar panels.

## 18.2   Pumped Storage Hydroelectricity System

Pumped hydroelectric plants use excess electricity during low demand and pump water from a lower altitude reservoir to an upper reservoir and store the energy as potential energy. During high demand, water is released from the reservoir to turn a water turbine and drive an electric generator to produce electricity. Both the energy and power densities of pumped hydro plants depend on the elevation difference of the upper and lower reservoir or the head of the water. Therefore, to maximize the capacity of a pumped hydro plant, the site must have a high variation in topographic elevation. The application of pumped hydroelectric plants is limited to large-scale (+100 MW) and to specific geographic locations. Due to the maturity and reliability of the system, pumped hydro is the primary energy storage system deployed in many countries.

## 18.3   Compressed Air Energy Storage System (CAES)

Similar to pumped hydro, compressed air energy storage (CAES) compresses ambient air during times of low demand and stores it under pressure for use at times of high demand. The compressed air is usually stored in an underground cavern. When electricity is needed, the pressurized air is heated and run through a turbine that is connected to a generator to produce electricity. Therefore, a CAES plant functions like a conventional gas turbine plant. Instead of using the turbine work to compress the gas the air is pre-compressed using other energy sources. However, CAES is mostly limited to artificially constructed salt caverns in deep salt formations, and therefore, only one 110 MW CAES plant exists in the United States in McIntosh, Alabama. Another limitation of a CAES plant is its use of fossil fuel to heat the air before it enters the turbine if there is no thermal storage system installed to capture the heat produced during the compression process.

## 18.4   Hydrogen Energy Storage System (HES)

Hydrogen storage technology falls into two broad categories. The first category, physical storage of the hydrogen molecule, is the most common. Physical storage includes compressed hydrogen, liquefied hydrogen, and combined compressed and cooled hydrogen. The second category is material-based storage of hydrogen atoms. Material-based storage includes hydrides, sorbents, and chemical storage. Physical storage remains the most mature and economical technology.

Compressed hydrogen is typically stored in all-steel cylinders (available in large quantities at typical prices of $4.00–$5.00 per liter of storage) or in composite-wrapped cylinders which are commercially available from several suppliers. The composite cylinders are ideal for applications where reduced weight is a design criterion, such as mobile applications, but costly due to their complex manufacturing and certification process. Prices for composite-wrapped cylinders range from $27.00 to $49.00 per liter of storage.

Liquid hydrogen storage requires cooling systems that are capable of maintaining temperatures below hydrogen's boiling point, $-252.882\,^{\circ}$C. Combined compressed/cooled hydrogen storage can be maintained at slightly higher temperatures because compression is used to raise the boiling point. On a volumetric energy density basis, liquefied hydrogen is competitive with compressed natural gas (CNG), but both storage methods require large amounts of energy and infrastructure investments. Any heat transferred to the hydrogen results in boil-off and venting, reducing the amount of usable fuel and time hydrogen can remain in liquid form without expending energy for cooling.

Material-based storage is implemented by bonding hydrogen with other sub-stances through the use of metal hydrides, sorbents, or chemical storage. Metal-hydride storage devices have been proven to work for long-term hydrogen storage but are heavy, contain rare and expensive materials, and typically require thermal management systems to absorb and release hydrogen.

The two cheapest systems are compressed gas storage and sorbent-based storage. The 700 bar storage systems cost roughly the same as the most advanced sorbent-based systems, approximately $15 per kilowatt hour or $54 per megajoule. The major suppliers of hydrogen tanks are Dynetek Industries Ltd. and Quantum Fuel Systems Technologies Worldwide, Inc., based in California. They quote a storage system cost of $10/kWh. The technology is quite mature for stationary applications, with more than 1 million high-pressure cylinders in operation worldwide. For transport applications there is a need for higher densities. Storage tanks at 700 bar made of composite materials have already entered the market. The compressed hydrogen is stored in 350 bar lightweight, aluminum-lined, carbon-fiber-reinforced cylinders (174 l water volume each), which are manufactured by Dynetek Industries Ltd. This company also supplies carbon composite cylinders rated to 875 bar that weigh only a third of equivalent steel cylinders. Dynetek model W290 tanks with a water volume of 290 liters measure 416 mm in diameter and 2880 mm in length. For a pressure of 250 bar, they store 80 kg of hydrogen at a weight of 121 kg. Quantum technologies produce a tank at a service pressure of 350 bar. It has a diameter of 274 mm and a length of 940 mm. It can store 1.55 kg of hydrogen at a water volume of 40 liters and weighs 20 kg. After delivery of the hydrogen on shore, the hydrogen can be stored in large tanks. During their wind/hydrogen demonstration project, the Norsk Hydro Company [3] used a 2400 Nm$^3$ (215.7 kg) hydrogen tank at 200 bar. In the Grimsey Island study [4], a 250 kg hydrogen storage tank costing $185,000 was considered for the wind-diesel-hydrogen scenario and a 850 kg hydrogen tank costing $630,000; hence, the storage cost per kg hydrogen is $740/kg.

## 18.5   Ammonia Storage and Transportation System

A promising additional way of storing and transporting hydrogen is to convert it into ammonia and reconverting it back into hydrogen at the point of end use for fuel cell vehicles or used directly in solid oxide fuel cells, in internal combustion engines or gas turbines. One kg of ammonia production requires 10–12 kWh of energy using either the Haber-Bosch process or electrochemical synthesis. Furthermore, its conversion back to hydrogen can lead to substantial losses. Nevertheless, ammonia appears to offer advantages over hydrogen for the transport of renewable energy from regions with high intensity to regions lean in renewable energy sources because the infrastructure for the transport of liquid hydrogen is still in its infancy. A round-trip efficiency (RTE) can be defined as the net useful energy for an end application from the total energy input at the point of ammonia production. More detailed analysis can be found in references [5, 6]. Recent reviews of large-scale storage of hydrogen and of ammonia-based hydrogen storage were published in references [7, 8].

# References

1. F. Ausfelder et al., Energy storage as part of a secure energy supply. ChemBioEng Rev. **4**(3), 144–210 (2017)
2. S. Watson, A. Frank, *A Fifteen Year Roadmap Toward Complete Energy Sustainability*. UC Davis Institute of Transportation Studies Research Report UCD-ITS-RR-12-35, January 2012
3. O. Ulleberg, T. Nakken, A. Ete, The wind-hydrogen demonstration system at Utsira in Norway: Evaluation of system performance using operational data and updated hydrogen energy system modeling tools. Int. J. Hydrog. Energy, 1–12 (2009)
4. D. Chade, T. Miklis, D. Dvorak, Feasibility study of wind-to-hydrogen system for Arctic remote locations – Grimsey island case study. Renew. Energy **76**, 204–211 (2015)
5. S. Giddey, S.P.S. Badwal, C. Munnings, M. Dolan, Ammonia as a renewable energy transportation media. ACS Sustain. Chem. Eng. **5**(11), 10231–10239 (2017)
6. W.C. Leighty, J.H. Holbrook, *Running the World on Renewables: Alternatives for Transmission and Low-cost Firming Storage of Stranded Renewables as Hydrogen and Ammonia Fuels via Underground Pipelines*. ASME Int. Mech. Eng. Congress, Houston, Texas, November 9–15, 2012, IMECE2012-87097
7. J. Andersson, S. Groenkvist, Large-scale storage of hydrogen. Int. J. Hydrog. Energy **44**(23), 11901–11919 (2019)
8. J.W. Makepeace et al., Reversible ammonia-based and liquid organic hydrogen carriers for high-density hydrogen storage. Int. J. Hydrog. Energy **44**(15), 7746–7767 (2019)

# Chapter 19
# Hydrogen Compression Technology

Most hydrogen compressors used today are mechanical diaphragm or piston compressors. Mechanical compression systems have relatively simple construction, maintenance, and repair procedures. Several major manufacturers offer mechanical compressors with a wide range of inlet and outlet pressure configurations, with and without integrated cooling, lubricated, or unlubricated. While the technology for mechanical compression is mature, they have limited compression ratios. Piston compressors are limited to a single-stage compression ratio of 4–6:1, while diaphragm compressors can achieve 15–20:1 ratios in a single stage. In contrast, electrochemical hydrogen compressors have demonstrated compression ratios of 300:1. Mechanical compressors are also expensive both in up-front capital expenditure requirements and operation and maintenance. Mechanical compressors are also large, heavy, and loud and, usually, require several hazardous materials to operate efficiently. "Small" mechanical compressors can weigh as much as 200–400 kg. The smallest mechanical compressor available on the market weighs 170 kg and requires $0.5 \text{ m}^3$, while it could only compress to 51 bar. Operating this compressor would require hearing protection and handling of hydraulic fluid and lubricants. Electrochemical hydrogen compressors, on the other hand, are silent and compact and do not require handling hazardous materials [1].

The theoretical energy to compress hydrogen isothermally from 20 bar to 350 bar is 1.05 kWh/kg $H_2$ and only 1.36 kWh/kg $H_2$ for 700 bar. The minimum theoretical energy to liquefy hydrogen from ambient conditions is 3.3 kWh/kg $LH_2$ or 3.9 kWh/ kg $LH_2$ with conversion to para-$LH_2$ (which is standard practice). Actual liquefaction energy requirements are substantially higher, typically 10–13 kWh/kg $LH_2$, depending on the size of the of liquefaction operation. Novel liquefaction methods, such as an active magnetic regenerative liquefier, may require as little as 7 kWh/kg $LH_2$. For comparison the lower heating value (LHV) of hydrogen is 33.3 kWh/kg $H_2$. Hydrogen is typically compressed by a reciprocal compressor which can achieve an isentropic efficiency of about 56% and a motor efficiency of 92%. Depending on the type of compressor used, 2–4 kWh/kg $H_2$ are usually quoted as needed to

M. F. Platzer, N. Sarigul-Klijn, *The Green Energy Ship Concept*, SpringerBriefs in
Applied Sciences and Technology, https://doi.org/10.1007/978-3-030-58244-9_19

compress hydrogen to 350 bar. The National Renewable Energy Laboratory quotes the compressor costs to be $600,000, $300,000, $100,000 in the near-, mid-, and long-term for the compression of 1500 kg of hydrogen to 6500 psi [2].

Electrochemical hydrogen compressors (EHCs) are solid-state devices that use direct current electricity to transport hydrogen through a proton exchange membrane and build pressure into a pressure vessel. Their physical construction, operation, and theory are very similar to that of a proton exchange membrane fuel cell. There are numerous potential advantages to using EHCs as opposed to traditional mechanical compressors; most notably, the solid-state EHCs are not subject to the same mechanical friction and thermodynamic losses of their mechanical counterparts. The EHC is also designed to follow an isothermal compression process which requires less energy than the adiabatic process of mechanical compressors. A third core advantage is the inherent purification process that happens as hydrogen gets transported through the membranes.

As low-pressure hydrogen is supplied to the inlet (anode), each hydrogen atom loses an electron at the anode, and this electron gets transported via the electrical power supply to the cathode. Since the former hydrogen atom is now missing an electron, it becomes a proton which is attracted to the cathode and pulled through the membrane. At the cathode, each proton receives an electron, becomes a hydrogen atom, bonds with another hydrogen atom, and exits through the compressor outlet. As hydrogen flows out of the compressor outlet, it fills the storage vessel and increases the vessel pressure until the power supply is turned off, a relief valve is opened, or the compressor reaches its maximum compression.

# References

1. M. Gardiner, *Energy Requirements for Hydrogen Gas Compression and Liquefaction as Related to Vehicle Storage Needs*. DOE Hydrogen and Fuel Cells Program Record, 7 July 2009
2. J. Levene, B. Kroposki, G. Sverdrup, *Wind Energy and Production of Hydrogen and Electricity – Opportunities for Renewable Hydrogen*. Conference Paper NREL/CP-560-39534, March 2006

# Chapter 20
# Power from Air and Water Flows

Two scenarios are of interest. We first consider a stationary wind or hydropower generator exposed to a wind or water stream. We want to estimate the maximum power which can be extracted from a wind or water stream of a given velocity. As is well-known, this can be done by using actuator disk theory assuming inviscid incompressible flow conditions.

Consider the actuator disk shown below. The actuator disk is a fictitious turbine consisting of infinitely many blades which, nevertheless, allow flow through the disk. Immediately upstream of the disk the pressure has the value $p$. The disk extracts a certain amount of power from the flow, causing the velocity $V$ to decrease by the amount $v$ and the pressure to decrease by the amount $p'$. Far upstream the pressure has the value $p_0$. Far downstream the pressure assumes again the value $p_0$, but the velocity decreases by the amount $v_1$. The Bernoulli equation may be applied upstream and downstream of the actuator disk. Hence, the total upstream pressure head is $H_0$. Therefore, we may write (Fig. 20.1):

$$H_0 = P_0 + \rho V^2/2 = P + \rho(V-v)^2/2 \qquad (20.1)$$

The total downstream pressure head is $H_1$. Therefore, we may write with $\rho$ as fluid density:

$$H_1 = P_0 + \rho(V-v_1)^2/2 = P - P' + \rho(V-v)^2/2 \qquad (20.2)$$

M. F. Platzer, N. Sarigul-Klijn, *The Green Energy Ship Concept*, SpringerBriefs in Applied Sciences and Technology, https://doi.org/10.1007/978-3-030-58244-9_20

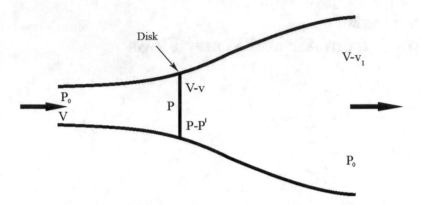

**Fig. 20.1** Actuator disk

and hence, the pressure jump across the disk is given by the difference in pressure heads:

$$P' = H_0 - H_1 = \rho V^2/2 - \rho/2\left[V^2 - 2Vv_1 + (v_1)^2\right]$$
$$= \rho(V - v_1/2)v_1 \tag{20.3}$$

The rate of decrease in axial momentum is the drag on the disk, hence with $A$ as disk area:

$$D = A\rho(V - v)v_1 \tag{20.4}$$

Since $P'$ is the drag per unit area of the disk, by equating the two expressions for $P'$, it follows that

$$\rho(V - v)v_1 = \rho(V - v_1/2)v_1$$
$$v = v_1/2$$

It is seen that the velocity decrease at the disk is half the decrease far downstream. The decrease of kinetic energy in unit time, i.e., the power output, is given by

$$P = A\rho/2(V - v)\left\{V^2 - (V - v_1)^2\right\} = A\rho/2(V - v)^2 4v$$

This power becomes a maximum for $v = V/3$ $\qquad$ (20.5)

Introducing the factor $a = v/V$ the extracted power can be written as

$$P = A(\rho/2)V^3(1 - a)^2 4a$$

and non-dimensionalizing by referring to the inflowing kinetic power one obtains the coefficient of performance $C_P = (1-a)^2\,4a$.

Another common way of writing this result is to introduce the axial induction factor.

$$\zeta = (V - 2v)/V \tag{20.6}$$

represents the ratio of the final downstream speed and the incoming flow speed. It then follows that

$$\zeta = 1 - 2a \tag{20.7}$$

and the coefficient of performance becomes

$$C_P = (1 + \zeta)(1 - \zeta^2)/2 \tag{20.8}$$

This coefficient becomes a maximum for

$$\zeta = 1/3 \tag{20.9}$$

The power of the air or water flowing through the power generator disk is $A\rho V^3/2$. Hence, the coefficient of performance represents the efficiency of the power generator by comparing the power output P with this quantity. The coefficient becomes a maximum for $\zeta = 1/3$ or $a = 1/3$ and assumes the value

$$\eta = 16/27 = 0.593 \tag{20.10}$$

This is the Betz limit [1], meaning that at most 59.3% of the available kinetic energy can be extracted from an incoming flow.

The second scenario applies for the case of a power generator which is moving relative to a stationary air or water environment. In this case, the turbine experiences a drag $D$. As shown by Glauert [2], this drag is found from the momentum decrease:

$$D = 2A\rho(V - v)v$$

and the power required to fly the turbine is given by $DV$, while the power which can be extracted from the turbine is given by

$$P = 2A\rho(V - v)^2v$$

The efficiency of the turbine then can be obtained by comparing the power output to the power needed to fly the power generator. Therefore,

$$\eta = P/DV = 2A\rho(V - v)^2v/2A\rho(V - v)v = (V - v)/V = 1 - v/V$$

The relationship between the efficiency and the power output is with d as disk diameter:

$$\eta^2(1 - \eta) = 2P/(\pi\rho V^2 d^2)$$

For a given flight speed and diameter, the power output is a maximum for $\eta = 2/3$ and $v/V = 1/3$.

Note that for a stationary turbine on the ground, the efficiency was defined by comparing the power output with the power flowing though the turbine, reaching a maximum value of 16/27 (59.3%) in contrast to 66% for the flying turbine but $v/V$ remaining 1/3.

At standard sea level conditions (15 degrees centigrade), the power density (power per unit disk area) for air is

$$P/A = \rho V^3/2 = 0.615\,V^3\,(\text{Watt}) \quad \text{if} \quad V(m/s)$$

and for water

$$P/A = 499.5\,V^3$$

Hence, for the same speed, the power density of water is larger by a factor of 812.

It is instructive to compare the disk areas required to generate 1 MW of shaft power at a wind or water speed of 10 m/s.

For air, $1,000,000 = A\,(0.593)\,(0.615)\,(10)^3$

$A = 2742\ \text{m}^2$

For water, $1,000,000 = A\,(0.593)\,(499.5)\,(10)^3$

$A = 3.38\ \text{m}^2$

For air this corresponds to a disk diameter of 59 m, whereas in water the disk diameter is only 2 m.

This rapid estimation of the power densities of air and water illuminates the size constraint encountered in the deployment of wind power generators. Typical average wind speeds are 10 m/s. Therefore, modern wind turbines, such as the Siemens SWT-6.0-154 turbine, have a rotor diameter of 154 m to produce 6 MW, corresponding to a rotor area of 18,626 $\text{m}^2$ and blade lengths of some 70 m, making their transport, erection, and maintenance a difficult and expensive operation. The size of hydropower generators, on the other hand, remains quite small provided water speeds of 3 m/s or more are available.

# References

1. A. Betz, *Das Maximum der theoretisch moeglichen Ausnutzung des Windes durch Windmotoren* (Zeitschrift fuer das gesamte Turbinenwesen, 1920), pp. 307–309
2. H. Glauert, *The Elements of Aerofoil and Airscrew Theory* (Cambridge University Press, 1926), pp. 206–207

# Chapter 21
# Hydrokinetic Turbine Technology

As is well-known from the Betz limit, at most only 59.3% of the kinetic energy flowing through an area A can be extracted. Therefore, if one wants to extract 1 kW of power from a water flow at 1.25 m/s requires an area of 1.727 m$^2$, corresponding to a turbine rotor diameter of 1.48 m. An increase of the water speed to 2 m/s reduces the required area to 0.42 m$^2$ and the rotor diameter to 0.73 m. The actual required areas are significantly larger depending on the turbine efficiency.

The basic physics of power extraction from flowing fluids therefore inevitably leads to large turbine sizes. Since the power is proportional to the third power of the speed, a doubling of the flow speed generates an area reduction by a factor of eight. Therefore, even small flow speed increases will pay off in terms of turbine size and cost.

The large required turbine size at low water speed implies not only a relatively large turbine cost but also difficulties in installing and maintaining the turbine. This aspect suggests a need to search for the right trade-off between turbine size and power output.

Very small hydrokinetic turbines with power outputs between 100 and 600 W are commercially available from companies, such as Watt and Sea and Ampair. In 2009 the Canadian company RER Hydro developed and installed a shrouded turbine with a 2.8 m rotor diameter and a maximum power output of 350 kW (at a flow speed of 4.5 m/s) in the St. Lawrence river. The diffuser exit area was 20.69 m$^2$, hence 3.6 times larger than the rotor area. The turbine was anchored close to the riverbed. After a successful testing period of several years, the project was terminated.

The Virginia company Verdant Power installed 15-foot-diameter unshrouded 36 kW turbines on the bottom of New York City's East River. Turbine failures caused by continuous operation in the harsh marine environment without easy access to the turbines for maintenance and repair are still preventing the proposed expansion of the system to 200 or 300 turbines.

M. F. Platzer, N. Sarigul-Klijn, *The Green Energy Ship Concept*, SpringerBriefs in
Applied Sciences and Technology, https://doi.org/10.1007/978-3-030-58244-9_21

The Austrian company Aqua Libra developed and tested a shrouded turbine in the Danube. It has a 2.5 m rotor diameter and a maximum power output of 70 kW at a flow speed of 3.4 m/s. The turbine delivers 6.3 kW at 1.3 m/s and 18.8 kW at 2 m/s. The diffuser exit area is 3.27 times larger than the rotor area. The turbine was anchored to the riverbed in such a way that it could float near the water surface.

As already noted, the efficiency of unshrouded turbines is limited to 59.3% by the Betz limit. This limit is based on the assumption that the free-stream is not disturbed by an external force prior to its interaction with the turbine. However, it is well recognized that altering the stream tube incident on the turbine rotor via some external mechanism can produce efficiencies in excess of the Betz limit. A shrouded wind turbine, often also called a diffuser augmented turbine, makes use of such a mechanism. Encasing the turbine in a shroud accelerates the oncoming flow, thereby significantly increasing the mass flow and the power extractable from the flow through the turbine. This consideration motivated the shrouding of the RER Hydro and Aqua Libra turbines. However, as also already noted, shrouding only reduces the size of the rotor but increases the overall size of the turbine because the diffuser exit area had to be more than three times larger than the rotor area for both turbines. In addition to the advantage of exceeding the Betz limit, shrouding offers another important advantage. Every turbine is limited by its so-called cut-in speed, a flow speed below which the blades do not turn. A shrouded turbine provides a lower cut-in speed than a conventional open rotor. This feature strongly favors the use of shrouded turbines in order to enable power extraction down to very low flow speeds, as typically encountered in rivers and tidal flows. An often cited disadvantage of shrouded turbines is the diffuser length required to avoid flow separation. Fortunately, shrouded turbines with much shortened diffusers have been demonstrated in recent years which achieve power augmentations by a factor of 2–5 compared with open rotors, for a given turbine diameter and wind speed. This is made possible by adding a broad-ring flange at the shortened diffuser exit, causing the formation of a strong vortex behind the flange and thus drawing more mass flow through the turbine rotor [1–3].

## 21.1  Turbine Design

The design methods used in the wind turbine industry can be applied on designing hydrokinetic turbines with some modifications to the design model. To this end, the well-known blade element theory and blade momentum theory are combined in the BEMT code, used by Turner at the University of Cincinnati [4]. Using BEMT, the power, the drag, and the geometric properties of the blade can be calculated. An iterative procedure is incorporated to obtain the axial and angular induction factors. This theory can be used to generate a blade as a starting point and can be made robust

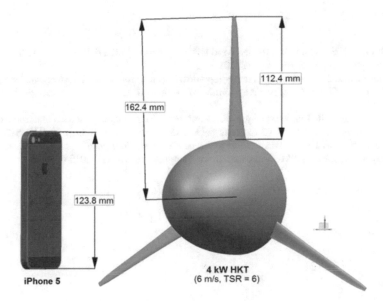

**Fig. 21.1**  Single-rotor hydrokinetic turbine [4]

using model corrections, computational fluid dynamics, and finite element analysis and shape optimization. The BEMT model is also made robust by accounting for the tip losses using a tip loss factor and by adding correction factors from the Glauert and Buhl empirical relationship. For the iterative blade design process, the University of Cincinnati in-house parametric 3D blade geometry builder (3DBGB) is used to create 3D blade shapes from the parameters obtained by the BEMT model. A combination of parameters in 3DBGB can be optimized for optimal blade shape for a wide range of operating conditions. The choice of airfoil is another crucial factor for the blade design along with choosing the range of angle of attack. The XFOIL code is used to make a robust choice of the airfoils, angle of attack range, and coefficient of lift and drag for the chosen airfoils. A 3D CFD analysis capability is also available to analyze the 3D effects of the blades. It is an expertise of the University of Cincinnati Gas Turbine Lab. It includes the use of Numeca's Fine/ Turbo of CD-Adapco's Star-CCM+ code.

Figure 21.1 shows an example design of K. Siddappaji and M. Turner [4] at the University of Cincinnati for a 5 kW single rotor hydrokinetic turbine with winglets, where they applied the above-described design capability for different configurations, such as unshrouded single row, unshrouded counter rotating, and shrouded nozzle-rotor-OGV configurations. They showed how this design and analysis system for hydrokinetic turbines can be linked to an optimizer to obtain optimal blade shapes, yielding an efficiency of 58.74% for a counter-rotating design. Note the small turbine size required for a 4 kW output.

# References

1. Y. Ohya, T. Karasudani, A shrouded wind turbine generating high output power with wind-lens technology. Energies **3**, 634–649 (2010)
2. A. Tourlidakis, K. Vafiadis, V. Andrianopoulos, I. Kalogeropoulos, *Aerodynamic Design and Analysis of a Flanged Diffuser Augmented Wind Turbine*. ASME Turbo Expo 2013, ASME-GT2013-95640
3. A.C. Aranake, V.K. Lakshminarayan, K. Duraisamy, Computational analysis of shrouded wind turbine configurations using a 3-dimensional RANS solver. Renew. Energy **75**, 818–812 (2015)
4. K. Siddappaji, M. Turner, *Revolutionary Geometries of Mobile Hydrokinetic Turbines for Wind Energy Applications*. ASME Int. Gas Turbine Congress, ASME-GT-2015-42342

# Chapter 22
# Wind-Propelled Ship Technology

## 22.1 Displacement Boats

A large amount of information is available about the design, performance, stability and control, and the operation of single or multiple-hull (catamaran or trimaran) conventional sailboats or sailing ships, for example, in Refs. [1, 2]. The ship drag arises from the hull friction drag and from the wave drag. The friction drag coefficient is obtained from the formula stipulated by the International Towing Tank Conference ITTC where the Froude number $\mathrm{Fr} = V/SQRT(gL)$ is used and where $V =$ ship speed (m/s), $L =$ ship length, and $g$ is the gravitational constant (m/s$^2$). where $R_f =$ viscous ship drag (N), $\rho =$ water density (kg/m$^2$), $S =$ wetted area (m$^2$) and $R_n =$ Reynolds number based on the DWL (design water line) (Table 22.1).

$$C_f = \frac{0.075}{(\log R_n - 2)^2} = \frac{R_f}{\frac{1}{2}\rho S V^2} \qquad (22.1)$$

The total ship drag coefficient then is obtained by adding the wave resistance coefficient.

Hongxuan Peng in his PhD thesis [3] provides computational estimates and comparisons with experiments for the wave resistance coefficient of a catamaran boat. He shows that for low Froude numbers between 0.2 and 0.4, the computed wave resistance coefficient varies between 0.001 and 0.002. The coefficient increases to 0.005 at Froude numbers of about 0.5 and decreases to values of 0.002 at Froude numbers of about 1.0. He shows good agreement between computed and experimental results for the Wrigley trimaran and for the Dalhousie trimaran. Furthermore, Peng also gave some comparisons of the wave drag coefficients for the wave cancellation multihull. His results indicate that wave resistance coefficients

M. F. Platzer, N. Sarigul-Klijn, *The Green Energy Ship Concept*, SpringerBriefs in
Applied Sciences and Technology, https://doi.org/10.1007/978-3-030-58244-9_22

**Table 22.1** Friction drag
coefficients for various
Reynolds numbers

| Reynolds number | Coefficient of friction |
|---|---|
| $10^5$ | 0.0083 |
| $10^6$ | 0.0047 |
| $10^7$ | 0.0030 |
| $10^8$ | 0.0021 |
| $10^9$ | 0.0015 |
| $10^{10}$ | 0.0012 |

**Fig. 22.1** Sail section
characteristics. (Source: [1,
page 390])

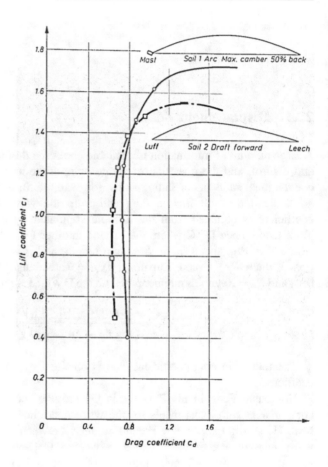

typically vary between 0.0005 and 0.002 at low Froude numbers and can reach
0.005 at Froude numbers between 0.3 and 0.6. At still higher Froude numbers, the
wave resistance coefficients decrease. Therefore, it appears justified to assume total
resistance coefficient variations between 0.004 and 0.008.

The total ship drag must be balanced by the sail thrust which is generated by one
or multiple soft, semirigid, or rigid sails. Figure 22.1 shows that a typical soft sail

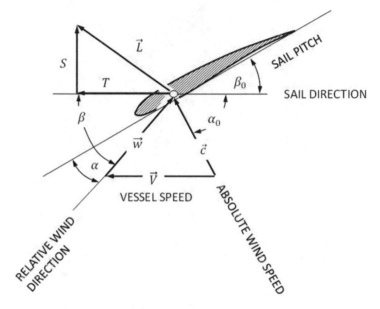

**Fig. 22.2**  Forces acting on the rigid sail of a sailboat [4]

generates a maximum lift coefficient of 1.7. As shown in Fig. 22.2, the total aerodynamic force L on the sail (represented by an airfoil) has a component T pointing forward which provides the thrust to move the boat forward in a specific direction. The other component S is the heeling force which tends to roll the boat and which therefore needs to be counteracted by a hydrodynamic force on the hull and keel of the boat. Sailing with the speed $V$ at an angle $\alpha_0$ to the absolute (true) wind speed $c$ exposes the boat to a relative (apparent) wind speed w at an angle $\beta$. It can be readily seen that the boat cannot sail directly into the wind but needs to tack alternately to the right and left in order to reach a point in the wind direction. Figure 22.3 shows the various wind angles at which the boat can sail. It is "running" when the boat sails with the wind and the apparent wind angle is 180°. It is sailing in the "broad reach" sailing mode when it is sailing at wind angles greater than 90°, in the "reaching" mode when the wind angle is approximately 90°, in the "close reaching" mode when it is sailing at wind angles less than 90°, and in the "close-hauled" mode when it tries to sail as close as possible against the wind.

   The use of sailing ships for power production by means of attached hydrokinetic turbines necessitates the maximization of the sail thrust and the minimization of the ship drag because the power production from the hydrokinetic turbines generates an unavoidable drag force, as shown in Chap. 20.

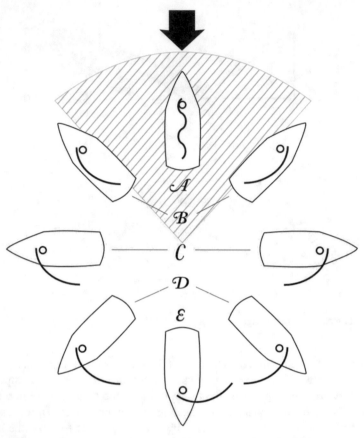

**Fig. 22.3** Sailing modes: *A* Into the wind (shaded "no go zone" where a craft may be "in irons). *B* Close-hauled; *C* beam reach; *D* broad reach; *E* running. (Source: Wikipedia, Points of Sail)

For this reason, it is crucial to determine the sailing mode which maximizes the sail thrust. As can be seen from Fig. 22.2, the thrust force is given as a trigonometric function of the lift force [4]:

$$T = \sin\beta \frac{\varrho_g}{2} A c_L(\alpha) w^2 \tag{22.2}$$

where $\beta$ denotes the relative wind direction and $w$ denotes the relative wind speed and $c_L$ the lift coefficient, which is a function of the angle of attack $\alpha$. The cosine rule gives a relation between the velocities at the sail section:

$$w^2 = V^2 + c^2 - 2Vc\cos\alpha_0 \tag{22.3}$$

where $c$ denotes the absolute wind velocity and $\alpha_0$ the absolute wind direction. The absolute wind velocity is the vector sum of the relative wind and the vessel speed

$\vec{c} = \vec{V} + \vec{w}$. By introducing the dimensionless ship speed $v := V/c$, the air density $\rho_g$, the dimensionless thrust is given by

$$Ct = \frac{T}{\frac{\rho_g}{2}c^2 A} = \sin\beta c_L \left(v^2 + 1 - 2v\cos\alpha_0\right).$$

or after re-expressing the above expression by a function of $\alpha_0$ only

$$Ct = c_l(\sin\alpha_0)\ SQRT\ \left(v^2 + 1 - 2v\cos\alpha_0\right)$$

or

$$Ct = c_l\ SQRT\ \left[(1 - \cos 2\alpha_0)\left(v^2 + 1 - 2v\cos\alpha_0\right)\right]$$

Differentiation with respect to $\cos\alpha_0$ and equating to zero then yields the absolute wind angle for maximum thrust as a function of the dimensionless ship speed, displayed in Fig. 22.4. It is seen that the absolute wind angle increases from $90°$ at zero ship speed to 106.90 at half wind speed and to 109.50 at full wind speed and then declines again toward $90°$ as the ship speed increases to multiples of the wind speed.

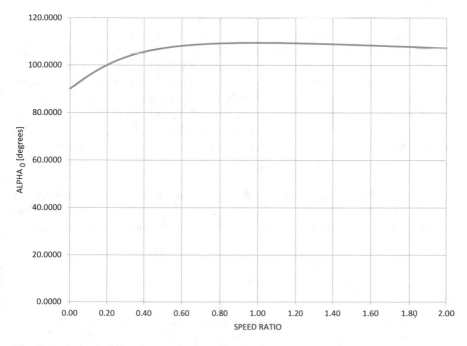

**Fig. 22.4** Optimal wind angle versus ship speed [5]

## 22.2   Hydrofoil Boats

The requirement to minimize the ship resistance leads to the use of hydrofoils in order to reduce the boat drag. The first suggestion to use hydrofoils can already be found in the British patent granted to Emmanual Denis Farcot in 1869, who claimed that "adapting to the sides and bottom of the vessel a series of inclined planes or wedge-formed pieces, which as the vessel is driven forward will have the effect of lifting it in the water and reducing the draught." The credit of first having demonstrated a hydrofoil boat on Lago Maggiore in 1906 belongs to the Italian Enrico Forlanini. A comprehensive history of the powered hydrofoil boats built in the last century was published by John R. Meyer, Jr. [6].

In 1972 the fastest sail boats could reach 26 knots. However, Bernard Smith [7], the former director of the Naval Weapons Laboratory in Dahlgren, Virginia, had already conceived a radically new design for a 40 knot sailboat in 1963 which consisted of an airfoil to provide lift and of a hydrofoil connected to the airfoil with a long boom to counteract the heeling force. This concept was taken up years later by Paul Larsen and Malcolm Barnsley, a Vestas wind turbine test engineer. With support from the Vestas company, they were eventually able to push their boat, the Sailrocket, to a speed of 40 knots in 2007. A major redesign of the boat, the Sailrocket2, finally made possible the all-time speed record of 65 knots (75 miles per hour) in 30 knot winds for a wind-powered boat in November 2012 over a 500 m course in Walvis Bay, Namibia [8].

The credit for this achievement belongs to Bernard Smith who gave the definition of an aerohydrofoil as "the machine that consists of two vertical wings, an inverted one in the water joined to an erect one in the air. When coupled in this way the assembly may be likened to a sailboat that has a sail and a centerboard, but no hull; except that the sail is no longer a conventional flexible sail but a rigid airfoil, and the centerboard no longer a centerboard but a hydrofoil" [7, page 3]. He recognized that stability could be achieved by connecting the airfoil and hydrofoil by a long boom. As a consequence, the boat sailing at 80° to the wind, in theory, can travel almost six times as fast as the wind. This technique relies on having enough resistance to stop the boat being pushed sideways by the wind. The drag from the hydrofoil determines the boat speed.

The sail rocket greatly stimulated the interest in hydrofoil sailing. Indeed, the America's Cup boats can reach speeds up to three times the speed of the wind. The top speed in the 34th America's Cup was 47.57 knots in a wind of 21.8 knots.

## References

1. C.A. Marchaj, *Aero-Hydrodynamics of Sailing* (Dodd, Mead & Company, New York, 1979)
2. F. Bethwaite, *High Performance Sailing* (Adlard Coles Nautical) 2nd edition reprinted 2014
3. H. Peng, *Numerical Computation of Multi-Hull Ship Resistance and Motion*, PhD thesis, Dalhousie University, June 2001

4. P.F. Pelz, M. Holl, M.F. Platzer, Analytical method towards an optimal energetic and economical wind-energy converter. J. Energy **94**, 344–351 (2016)
5. M.F. Platzer, N. Sarigul-Klijn, Storable energy production from wind over water. J. Energy Power Technol. **2**(2) (2020). https://doi.org/10.21926/jept.2002005
6. J.R. Meyer Jr., *Ships that Fly, a Story of the Modern Hydrofoil* (Hydrofoil Technology Inc., 1990)
7. B. Smith, *The 40-Knot Sailboat* (Echo Point Books & Media, LLC, 1963)
8. J. Winters, Will the sailrocket usher in a new age of sail? Mech. Eng. Mag. ASME **135**(8), 48–55 (2013)

# Chapter 23
# Power from Wind Over Water

The winds over water surfaces, ocean or lake waters, can be converted into electric energy using wind-powered vessels whereby hydrokinetic turbines are used to convert the kinetic energy of the water flow relative to the sailing vessel into electricity. In a further step, the electricity is stored in electric batteries or is used to produce hydrogen by electrolytically splitting sea water into hydrogen. The new feature of this "energy ship concept" is the shift of the fluid from air $(g)$ to water $(l)$ for the turbine. While the vessel is driven by wind power, the hydraulic turbine is driven by the water current. For the same turbine shaft power, the diameter ratio $D_g/D_l$ scales with the density ratio $\varrho_l/\varrho_g$:

$$\frac{D_g}{D_l} = \sqrt{\frac{\varrho_l}{\varrho_g}} \approx 3.5\%. \tag{23.1}$$

Therefore, the diameter of a hydrokinetic water turbine measures only 3.5% of a wind turbine, converting the same amount of energy. The smaller diameter of the turbine reduces the investment costs, which scale with the component size. The converted electricity needs to be stored using an adequate energy storage technology like storage in electric batteries or conversion into hydrogen by means of electrolysis. The hydrogen is then stored in tanks in the hull of the vessel. For ship propulsion either conventional soft sails, semirigid or rigid wing sails can be used, but controlled parafoils can also be employed to replace or complement the conventional or rigid wing sails in order to capture a greater amount of wind energy due to the greater height above the ocean surface.

Separating the energy converter into functional units these units are shown in Fig. 23.1. The two emerging subsystems are the vessel including the propulsion system and the hydrokinetic turbine which is described through the actuator disk theory [20.1].

Without loss of generality, a rigid sail is considered as vessel propulsion creating the thrust force $T$. The thrust is balanced by both the hydrodynamic drag force of the

M. F. Platzer, N. Sarigul-Klijn, *The Green Energy Ship Concept*, SpringerBriefs in
Applied Sciences and Technology, https://doi.org/10.1007/978-3-030-58244-9_23

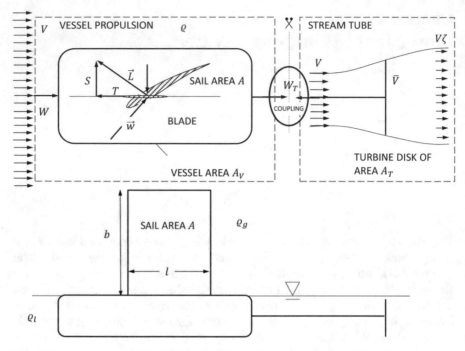

**Fig. 23.1** Representation of the energy ship [1]

vessel $W = A_V c_D V^2 \varrho_l / 2$ and the drag force of the turbine $W_T$, leading to the force balance $T = W + W_T$. The wetted vessel area is denoted as $A_V$, the turbine area as $A_T$, and the sail area as $A$. The drag coefficient of the vessel is denoted as $c_D$ and the vessel speed as $V$. The drag force of the turbine is given by the momentum balance for the turbine $W_T = \Delta p \, A_T$ with the pressure difference determined by Bernoulli's equation $\Delta p = V^2 (1 - \zeta^2) \varrho_l / 2$. The axial induction factor $\zeta$ is defined as the ratio of the downstream and upstream velocity (see Eq. 20.6) and is thereby a quantity for the flow deceleration through the turbine. The turbine is assumed to be sufficiently submerged so that the influence of the free surface can be neglected. Also side forces caused by the sail and turbine are neglected. Figure 23.1 shows the vessel and wind speed as well as the forces on a sail section.

Neglecting friction and induced drag for the first approach, the thrust force is given as a trigonometric function of the lift force:

$$T = \sin \beta \frac{\varrho_g}{2} A c_L(\alpha) w^2 \tag{23.2}$$

$\beta$ denotes the relative wind direction as shown in Fig. 22.2. $w$ denotes the relative wind speed and $c_L$ the lift coefficient, which is a function of the angle of attack $\alpha$. The cosine rule gives a relation between the velocities at the sail section:

$$w^2 = V^2 + c^2 - 2Vc \cos \alpha_0 \tag{23.3}$$

where $c$ denotes the absolute wind velocity and $\alpha_0$ the absolute wind direction. The absolute wind velocity is the vector sum of the relative wind and the vessel speed $\vec{c} = \vec{V} + \vec{w}$. By introducing the dimensionless absolute velocity $v := V/c$, the dimensionless air density $\varrho := \varrho_g/\varrho_l$, the dimensionless thrust is given by

$$\frac{T}{\frac{\varrho_l}{2} c^2 A} = \sin \beta c_{L\varrho}(v^2 + 1 - 2v \cos \alpha_0). \tag{23.4}$$

Then the dimensionless force balance is obtained:

$$\sin \beta c_{L\varrho}(v^2 + 1 - 2v \cos \alpha_0) = v^2 c_D a_V + v^2(1 - \zeta^2)a_T, \tag{23.5}$$

yielding a relation between the dimensionless vessel speed and the axial induction due to the turbine. Here the dimensionless areas are $a_V := A_V/A$ and $a_T := A_T/A$. It can be rewritten as

$$v^2 - 2vq \cos \alpha_0 + q = 0, \tag{23.6}$$

where

$$q := \frac{\sin \beta \, c_{L\varrho}}{\sin \beta \, c_{L\varrho} - c_D a_V - (1 - \zeta^2)a_T} < 0. \tag{23.7}$$

Hence, the dimensionless vessel speed is

$$v_{1/2, \, F} = q \cos \alpha_0 \pm \sqrt{q^2 \cos^2 \alpha_0 - q}, \tag{23.8}$$

yielding the relationship between the dimensionless vessel speed and all influencing physical parameters. Only the positive solution for the dimensionless vessel speed is physically reasonable. The negative solution represents the dimensionless vessel speed in inverted direction. The advantage of using dimensionless quantities is obvious: the amount of physical parameters is reduced so that a simpler equation to describe the system is obtained. Because Eq. (23.8) is a result of the force balance, it is denoted as $v_F$. Through this description the vessel speed is a function of seven parameters:

$$v_F = fn(\alpha_0, \beta, c_L, c_D, a_V, a_T, \zeta), \tag{23.9}$$

containing the approach flow, the shape and quality of the vessel and sail, the geometric areas, and the operation of the turbine. Only six of them are independent parameters. The relative wind direction will adjust in dependence of the vessel speed. Another equation, which can be found from the velocity triangle of Fig. 22.2, has to capture this relation. By applying the basic triangle equations for the velocities, we can find for the vessel speed $V$:

$$V = \cos \alpha_0 \, c + \cos \beta \, w. \tag{23.10}$$

With the velocity ratio $w/c = \sin \alpha_0 / \sin \beta$ the dimensionless vessel speed is

$$v_V = \cos \alpha_0 + \cot \beta \sin \alpha_0. \tag{23.11}$$

This dimensionless vessel speed is denoted as $v_V$, because it results from the velocity triangle at the sail section. The required equality of these two calculated dimensionless vessel speeds $v_F \overset{!}{=} v_V$ leads to an expression for the axial induction factor $\zeta$ as a function of the relative wind direction:

$$\zeta = \sqrt{\frac{\sin \beta c_{L\varrho}}{a_T}\left(\frac{2v_V \cos \alpha_0 - 1}{v_V^2} - 1\right) + \frac{c_D a_V}{a_T} + 1}. \tag{23.12}$$

For maximal turbine resistance ($\zeta = 0$), the maximal angle $\beta$ will adjust to the minimal vessel speed. For vanishing turbine drag and power output ($\zeta = 1$), the relative wind direction $\beta$ will minimize, representing the maximal vessel speed for the current approach flow. The permissible solution space of $\zeta$ is limited to $0 < \zeta \leq 1$ *and* ; the vessel speed is a function of the six independent parameters:

$$v = fn(\alpha_0, c_L, c_D, a_V, a_T, \zeta). \tag{23.13}$$

In Chap. 20 it was shown that the energetic efficiency of a pure wind or water turbine is obtained from the coefficient of performance:

$$C_P = (1 + \zeta)(1 - \zeta^2)/2$$

In contrast, the energetic efficiency of the energy ship has to be obtained from the coefficient of performance $C_P := P_S/P_{avail}$, defined as the ratio of the turbine power and the available power. The turbine power is $P_S = \eta \Delta p A_T \overline{V}$ with the arithmetic average velocity $\overline{V} = V(1 + \zeta)/2$.

The turbine power therefore can also be written as $\rho_1 A_t V^3/2 \, (1 + \zeta)(1 - \zeta^2)/2$ and the available power as $\rho_g A \, c^3/2$.

The ratio of these two quantities, written non-dimensionally, therefore yields the coefficient of performance of the energy ship as

$$\frac{C_P}{\eta_T} = \frac{1}{\varrho} v^3 a_T \frac{1+\zeta}{2} \left(1 - \zeta^2\right).\tag{23.14}$$

To determine the optimal energy conversion system, the following conditions need to be satisfied:

$$\nabla\left(\frac{C_P}{\eta_T}\right) = \begin{pmatrix} \dfrac{\partial}{\partial a_0}\left(\dfrac{C_P}{\eta_T}\right) \\[2mm] \dfrac{\partial}{\partial c_L}\left(\dfrac{C_P}{\eta_T}\right) \\[2mm] \dfrac{\partial}{\partial c_D}\left(\dfrac{C_P}{\eta_T}\right) \\[2mm] \dfrac{\partial}{\partial a_T}\left(\dfrac{C_P}{\eta_T}\right) \\[2mm] \dfrac{\partial}{\partial a_V}\left(\dfrac{C_P}{\eta_T}\right) \\[2mm] \dfrac{\partial}{\partial \zeta}\left(\dfrac{C_P}{\eta_T}\right) \end{pmatrix} = 0.\tag{23.15}$$

The last condition addresses the optimal axial induction factor $\zeta_{opt}$, which is adjusted by the turbine control through the generator.

It is instructive to take a closer look at each term in the force balance Eq. (23.5):

$$\sin\beta c_{L}\varrho\left(v^2 + 1 - 2v\cos\alpha_0\right) = v^2 c_D a_V + v^2\left(1 - \zeta^2\right)a_T,$$

Dividing by $\rho$ and replacing the axial induction factor $\zeta$ by the non-dimensional velocity ratio $a = v/V$ and noting the relation $\zeta = 1 - 2a$, it reads

$$\sin\beta c_L\left(v^2 + 1 - 2v\cos\alpha_0\right) = \left(\frac{1}{\rho}\right)v^2 c_D a_V + \left(\frac{4}{\rho}\right)v^2\, a(1 - a))a_T,$$

The term on the left-hand side represents the thrust coefficient. The first term on the right-hand side is the friction coefficient. All drag contributions except the turbine drag are lumped into this term, i.e., the induced drag of the sail or wing, the hull drag of the ship, the drag of the foils needed for pitch and heel stability, and the drag of the hydrofoils in case of hydrofoil boats. The second term on the right-hand side represents the turbine drag.

It is seen from Fig. 23.2 that the thrust increases only slightly with increasing boat speed (expressed as the boat-to-wind speed ratio), whereas the friction drag increases more quickly. The maximum speed of the boat without turbine therefore is obtained

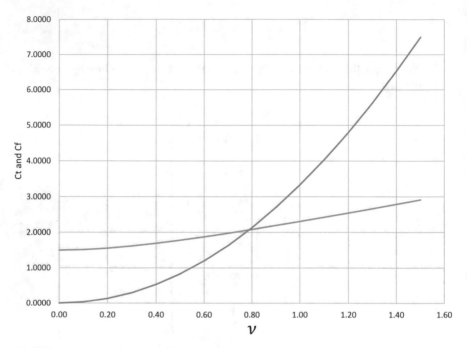

**Fig. 23.2** Thrust and friction coefficient versus speed ratio [23.4]

by the crossing of the two curves which occurs at a speed ratio of 0.8 for a wetted area ratio of 0.4 and a friction drag coefficient of 0.01.

Of special interest is the study of the influence of turbine drag on the turbine power output. It is well-known that the power output from stationary turbines is a maximum for $\zeta = 1/3$ and a = 1/3 (see Chap. 20). In contrast, the amount of power extraction from a moving turbine must be adjusted in such a way that the boat can move at the optimum speed. This is demonstrated in Fig. 23.3. Here the sum of the friction and the turbine drag are subtracted from the boat thrust. Force balance occurs at the crossing with the x-axis. It is seen that the speed is reduced as the turbine drag is increased from values of a = 0.025 to 0.05, 0.1, and 0.15. These values are far below the optimum a = 1/3 value for stationary turbines. Inserting these values into the equation for the coefficient of performance by rewriting Eq. (23.14) as

$$C_p \; = (4/\rho) \, a_t \, \nu^3 \, (1-a)^2 \, a$$

shows that the maximum power output value is reached for a = 0.1 and then starts to decrease for further decreases in a-values. This finding is in agreement with the optimum value of $\zeta = 0.81$ found in reference [1]. The reason for the much smaller a-value for maximum power production can be seen from the equation for the coefficient of performance which contains the third power of the speed ratio. The selection of larger a-values will cause a reduction in boat speed!

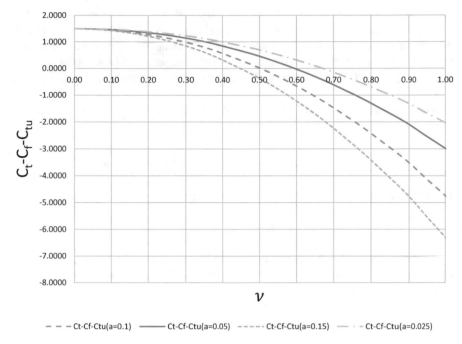

**Fig. 23.3** Force balance as a function of a-value [23.4]

The turbine power output P is obtained from the equation

$$P = \tfrac{1}{2}\, \rho_g\, c^3\, A\, C_p$$

For a boat with a wetted area ratio $a_V = 0.4$ and a turbine area ratio $a_T = 0.0124$, the optimal coefficient of performance is found to be $C_p = 0.4185$ at a speed ratio of 0.5. Therefore, the power output of a boat with a sail area of $A = 7$ m² sailing at optimal wind angle in a wind of 10 m/s becomes 1757.7 watt. This output reduces to 219.7 watt in a wind of 5 m/s. A boat with a sail area of 50 m² sailing in a wind of 10 m/s, on the other hand, generates 13 kW (in agreement with the value obtained in reference 23.1). The power output is proportional to the sail area. Hence, a ship with a sail area of 3200 m² can be expected to generate about 1 MW. The assumed values of wetted area and turbine area ratios are likely to be conservative. Hydrofoil-borne ships can be expected to generate double or even triple this output.

The available wind power is given by the circle whose diameter is the wing span. As shown by Prandtl [2], the wing lift can be written as

$$L = 2w\rho U\pi b^2 / 4$$

where w is the induced downwash velocity, U the free-stream-velocity, and b the wing span. In other words, the wing lift is equal to the mass flow through a circular air capture area of diameter b times twice the induced downwash velocity. The available kinetic energy per unit time therefore is given by $(1/2)\,\rho U^3 \pi b^2/4$ where $\pi b^2/4$ is the air capture area. Another important result obtained by Prandtl is the inverse relationship between the induced drag and the wing aspect ratio ($AR = b^2/S$, where S is the wing area). The ratio between the air capture area and the wing (or sail) then is $\pi\,AR/4$.

The efficiency of converting the available kinetic wind power into mechanical power is given by the ratio

$$\eta = P/P_{available} = \tfrac{1}{2}\,\rho_g\,c^3\,A\,C_p / \tfrac{1}{2}\,\rho_g\,c^3\,A_{capture}$$

which reduces to

$$\eta = 4\,C_p / \pi\,AR$$

Minimization of the induced drag requires large aspect ratio. Hence, the conversion efficiency of a boat with a turbine ratio $a_T = 0.0124$, a wetted area ratio of 0.4 and a wing with an aspect ratio of 10, will be

$$\eta = 4\,(0.4185)/10\,\pi = 0.0533$$

and for an aspect ratio of 4 eta is 0.133.

This conversion efficiency is certainly quite low compared to the efficiency of a stationary wind turbine. As shown in Chap. 20, the generation of 1 MW requires a minimum disk area of 2742 $m^2$ for the wind turbine in contrast to the sail (or wing) area of at least 3200 $m^2$ required by the energy ship. However, one advantage offered by the energy ship is the replacement of huge rotors by fixed wings. Additional advantages of the energy ship concept are the differences in capacity factor and increases in ship speed by means of hydrofoiling. It is likely that energy ships can reach capacity factors of 80% or even 90% due to their ability to operate year-round in high-wind ocean areas in contrast to 35–50% for land or offshore wind turbines. Hydrofoiling has the potential of doubling or even tripling the power output of displacement ships. For this reason, the authors are now developing and testing small-scale autonomous hydrofoil boats to serve as technology demonstrators for large-scale hydrofoil boats [3, 4].

In order to obtain the total combined mechanical and economic efficiency of the energy ship, the energy storage component has to be added. The turbine power is provided to the electrolyzer. The chemical power of hydrogen can be calculated as $\dot{m}_{H_2} H_{u,H_2}$, where $\dot{m}_{H_2}$ denotes the mass flow of hydrogen and $H_{u,H_2}$ the caloric value of hydrogen. To describe the system economically, the investment costs of all system components must be considered. They consist of the investment costs of the vessel $I_V$, the turbine $I_T$, and the storage technology $I_{st}$. Operation and maintenance costs are neglected in this first approach. The economical revenue can be calculated as the amount of hydrogen that can be sold.

To evaluate the system economically, the system is described through the economic profit function:

$$G = R - C.$$

$G$ denotes the economic profit, $R$ the revenue, and $C$ the investment costs. The energetic optimization yields the optimal axial induction factor, and the economical optimization yields the vessel and turbine dimensions. Additional details are described in references [1, 5, 6].

Very recently, Babarit et al. [7, 8] published a techno-economic assessment of the energy ship concept based on ships driven by Flettner rotors rather than sails and converting the wind energy into methanol. They found that the conversion efficiency is 24%.

# References

1. P.F. Pelz, M. Holl, M.F. Platzer, Analytical method towards an optimal energetic and economical wind-energy converter. J. Energy **94**, 344–351 (2016)
2. L. Prandtl, O.G. Tietjens, *Fundamentals of Hydro and Aerodynamics* (McGrawHill, New York/London, 1934)
3. M.F. Platzer, N. Sarigul-Klijn, *Mobile Offshore Platforms for Power Generation: The Energy Ship,* IOWTC2018-1022, ASME 2018 1st International Offshore Wind Technical Conference, San Francisco, November 4–7, 2018
4. M.F. Platzer, N. Sarigul-Klijn, Storable energy production from wind over water. J. Energy Power Technol. **2**(2), 1–12 (2020). https://doi.org/10.21926/jept.2002005
5. M. Holl, L. Rausch, P.F. Pelz, New methods for new systems – How to find the techno-economically optimal hydrogen conversion system. Int. J. Hydrogen Energy **42**, 22641–22654 (2017)
6. A. Ostolski, *Energieversorgungsoptimierung zur Potentialabschaetzung eines neuartigen Windenergiekonverters,* Forschungsbericht Fluidsystemtechnik, Technical University Darmstadt, S216, 23 November 2015
7. A. Babarit, S. Deloye, G. Clodic, J.C. Gilloteaux, Exploitation of the far-offshore wind energy resource by fleets of energy ships. Part A: Energy ship design and performance. Part B Cost Energy (2020). https://doi.org/10.5194/wes-2019-101
8. A. Babarit, G. Clodic, S. Deloye, J.C. Gilloteaux, Exploitation of the far-offshore wind energy resource by fleets of energy ships. Part A: Energy ship design and performance. Wind Energy Sci. (2020). https://doi.org/10.5194/wes-5-839-2020

# Chapter 24
# Conversion of Hydrogen to Electricity

A hydrogen fuel cell is an electrolyzer in reverse where hydrogen and oxygen are combined to produce electrical energy. The basic operations of a fuel cell are shown in Fig. 24.1. At the anode, the hydrogen gas ionizes and releases electrons. Because the electrolyte is ionically conducting and electrically insulating, the hydrogen ions pass through the electrolyte to the cathode, while the electrons pass through the external circuit when a load is connected. On the cathode side, the hydrogen ions recombine with electrons and oxygen to produce water and heat. For every 2 moles of water formed, 4 moles of electrons are produced. Because no combustion occurs in a fuel cell, its efficiency is not limited by Carnot efficiency and can ideally achieve 100% efficiency. However, inherent losses in the electrochemical reaction in a fuel cell reduce fuel cell efficiency. Different types of hydrogen fuel cells have power outputs ranging from watts to megawatts. For example, the company HORIZON produces a compact 3KW hydrogen fuel cell system which weighs only 11 Kg and has a size of 38 cm × 16 cm × 28 cm.

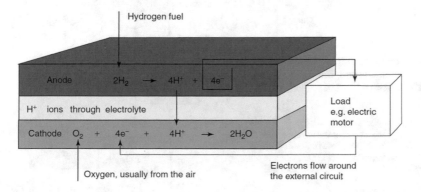

**Fig. 24.1** Fuel cell

M. F. Platzer, N. Sarigul-Klijn, *The Green Energy Ship Concept*, SpringerBriefs in
Applied Sciences and Technology, https://doi.org/10.1007/978-3-030-58244-9_24

## 24.1   Hydrogen Power Plant

In 2010 Jericha et al. [1] proposed to adapt the original Graz cycle for the burning of fossil fuels to hydrogen combustion with pure oxygen so that a working fluid of nearly pure steam becomes available. They showed that the cycle is able to reach a net cycle efficiency of 68% based on the lower heating value. The power plant incorporates solid oxide fuel cells into an innovative power cycle with steam as working fluid. As in the original cycle, it uses a high-temperature and a low-temperature cycle. The high-temperature part of the power plant consists of fuel cells, a combustion chamber, a high-temperature turbine, a high-pressure turbine, a compressor, and a heat recovery steam generator. The low-temperature steam loop consists of a low-temperature steam turbine, a condenser, a feed pump, a deaerator, and the steam supply to the steam compressor feeding the fuel cells. The basic layout of this hybrid Graz cycle is shown in Fig. 24.2. Steam is compressed to 41 bar and 600 degree Celsius starting from 1 bar at 100 degree Celsius, supplying 12 fuel cells arranged in parallel. Hydrogen and oxygen are fed in stoichiometric ratio into the fuel cells to deliver an output of 30 MW and to heat the steam to 800 degree Celsius. Additional hydrogen and oxygen then are burned in a combustion chamber behind the fuel cells, and the resulting hot steam is expanded through the high-temperature turbine to deliver an output of 123 MW. Further details are described in references [2, 3].

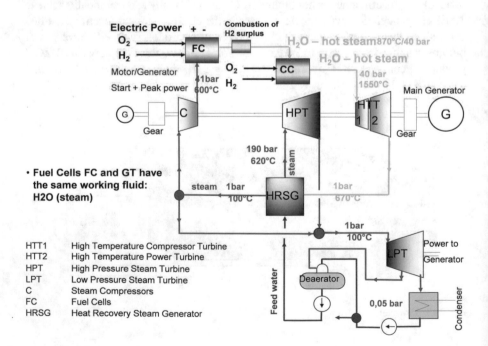

**Fig. 24.2**  Graz cycle power plant [1]

# References

1. H. Jericha, V. Hacker, W. Sanz, G. Zotter, *Thermal Steam Power Plant Fired by Hydrogen and Oxygen in Stoichiometric Ratio, Using Fuel Cells and Gas Turbine Cycle Components*, ASME GT2010-22282, June 14–18, 2010
2. M.F. Platzer, W. Sanz, H. Jericha, *Renewable Power Via Energy Ship and Graz Cycle*, 15th International Symposium on Transport Phenomena and Dynamics of Rotating Machinery, ISROMAC-15, February 24–28, 2014, Honolulu, Hawaii
3. W. Sanz, M. Braun, H. Jericha, M.F. Platzer, Adapting the Graz cycle for hydrogen combustion and investigation of its part load behavior. Int. J. Hydrog. Energy **43**(11), 5737–5746 (2018)

# Chapter 25
# Production of Jet Fuel from Seawater

The oceans contain approximately 100 mg of $CO_2$ per liter of seawater. Approximately 2–3% of this $CO_2$ is in the form of a dissolved gas, and the remaining 97% to 98% is in a chemically bound state as bicarbonate and carbonate. On a weight per volume basis, comparison extraction of $CO_2$ from the seawater instead of from the atmosphere is advantageous because of the higher water density. The simultaneous extraction of hydrogen and carbon dioxide then makes it possible to synthesize them into jet fuel based on Fischer-Tropsch technology provided a sufficiently abundant electric power source can be found for its implementation. Such a process not only is carbon-neutral but also eliminates the emission of sulfur and nitrogen compounds that are emitted during the combustion of petroleum-derived fuels.

Around the turn of the century the US Naval Research Laboratory initiated a comprehensive study of the technical and economic feasibility of producing jet fuel at sea using carbon dioxide and hydrogen because it can provide the US Navy significant logistical, operational, and cost advantages while implementing carbon-neutral aviation.

In 2012 Willauer et al. [1] concluded that the production of 100,000 gallons per day of jet fuel 443,900 $m^3$/day of $CO_2$ and 1,373,600 $m^3$/day of hydrogen is required. The production of this amount of hydrogen by large-scale commercially available alkaline or PEM electrolysers requires the processing of 1082 $m^3$/day of water and the delivery of 246 MW of electric power. The generation of the required electric power is by far the greatest capital cost. Willauer et al. [1] postulated the feasibility of building an ocean thermal energy conversion (OTEC) power plant at an estimated cost of $ 0.9 to $ 1.5 billion for a 200 MW power plant. Also, they estimated the additional costs for hydrogen production, carbon capture, and the production of jet fuel using the Fischer-Tropsch process as $194 million, $16 million, and $140 million for a total cost of $1.85 billion and $1.25 billion, respectively, producing 82,000 gallons of hydrocarbon fuel per day. This amounts to a cost of $8.70/gal or $5.78/gal. On the other hand, using either naval nuclear reactors or commercially available nuclear power plants to produce electric power the total

M. F. Platzer, N. Sarigul-Klijn, *The Green Energy Ship Concept*, SpringerBriefs in
Applied Sciences and Technology, https://doi.org/10.1007/978-3-030-58244-9_25

capital costs, they estimated a cost of \$1.35 billion and \$1.24 billion, respectively. Accounting for differences in amortization and operation/maintenance costs for commercial and naval reactors, the cost per gallon then comes to \$5.74/gal and \$2.90/gal, respectively.

In its most recent papers [2–4], the NRL team documented the development of an electrolytic cation exchange module for the simultaneous extraction of carbon dioxide and hydrogen gas from seawater. Work is currently underway to reach the goal of producing carbon dioxide and hydrogen in quantities and at efficiencies needed for the economically competitive production of synthetic jet fuels.

# References

1. Willauer, H.D., Hardy, D.R., Schultz, K.R., Williams, F.W., The feasibility and current estimated capital costs of producing jet fuel at sea using carbon dioxide and hydrogen, J. Renewable Sustainable Energy, Vo. 4, 033111 (2012). https://doi.org/10.1063/1.4719723
2. H.D. Willauer, F. DiMascio, D.R. Hardy, *Extraction of Carbon Dioxide and Hydrogen from Seawater by an Electrolytic Cation Exchange module (E-CEM) Part V: E-CEM Effluent Discharge Composition as a Function of Electrode Water Composition,* Naval Research Laboratory NRL/MR/6360-17-9743, August 1, 2017
3. F. DiMascio, D.R. Hardy, M.K. Lewis, H.D. Willauer, F.W. Williams, *Extraction of Carbon Dioxide and Hydrogen from Seawater and Hydrocarbon Production Therefrom,* U.S. Patent 9,303,323, April 5, 2016
4. H.D. Willauer, F. DiMascio, D.R. Hardy, F.W. Williams, Development of an electrolytic cation exchange module for the simultaneous extraction of carbon dioxide and hydrogen gas from natural seawater. Energy Fuel **31**, 1723–1730 (2017)

# Index